P9-DWY-133

DATE DUE

JN 2 8 99			

DEMCO 38-296

Science at the Bar

A TWENTIETH CENTURY FUND BOOK

The Twentieth Century Fund sponsors and supervises timely analyses of economic policy, foreign affairs, and domestic political issues. Not-for-profit and nonpartisan, the Fund was founded in 1919 and endowed by Edward A. Filene.

BOARD OF TRUSTEES OF THE TWENTIETH CENTURY FUND

Morris B. Abram, *Emeritus*
H. Brandt Ayers
Peter A. A. Berle
Alan Brinkley
José A. Cabranes
Joseph A. Califano Jr.
Alexander Morgan Capron
Hodding Carter III
Edward E. David Jr., *Emeritus*
Brewster C. Denny, *Emeritus*
Charles V. Hamilton
August Heckscher, *Emeritus*
Matina S. Horner
Lewis B. Kaden

James A. Leach
Richard C. Leone, *ex officio*
P. Michael Pitfield
Don K. Price, *Emeritus*
Richard Ravitch
Arthur M. Schlesinger Jr., *Emeritus*
Harvey I. Sloane, M.D.
Theodore C. Sorensen, *Chairman*
James Tobin, *Emeritus*
David B. Truman, *Emeritus*
Shirley Williams
William Julius Wilson

Richard C. Leone, *President*

К

Science at the Bar

Law, Science, and Technology in America

Sheila Jasanoff

A Twentieth Century Fund Book

Harvard University Press
Cambridge, Massachusetts, and London, England
1995

Riverside Community College
Library
AUG '96   4800 Magnolia Avenue
Riverside, California 92506

K 487 .S3 J37 1995

Jasanoff, Sheila.

Science at the bar

Copyright © 1995 by the Twentieth Century Fund, Inc.
All rights reserved
Printed in the United States of America

Library of Congress Cataloging-in-Publication Data
Jasanoff, Sheila
    Science at the bar: law, science, and technology in America /
Sheila Jasanoff.
        p. cm.
"A Twentieth Century Fund book."
Includes bibliographical references and index.
ISBN 0-674-79302-1
1. Science and law. 2. Technology and law. I. Title.
K487.S3J37 1995
344.73'09—dc20
[347.30495]   95–9192
                        CIP

For Maya

Contents

Foreword

Perhaps it is no surprise that in a society as diverse as America, the law and the courts play so large a role. These institutions in effect provide the mechanisms for sorting through the competing claims and conflicting ideologies of a uniquely complicated nation. Today the legal system routinely resolves questions that literally are beyond anything that could have been imagined when the foundations of the American legal system were set in place. Still, like other public institutions created by the English colonists, the courts have proven remarkable for their resiliency and their adaptability. Indeed, it is not too much to assert that the judicial system and the law have provided the framework for the orderly adaptation of many aspects of life to modernization, political evolution, and cultural change. But nothing tests the courts' capacity for resolving disputes and setting rules more than the legal issues raised by the scientific revolution.

It is not necessary to go back to the framers of the Constitution to locate a gulf in imagination and understanding concerning the explosion in scientific knowledge. The questions science puts before us today are well beyond the comprehension of most of us. We live, for the first time in the history of mankind, in a world in which the great mass of people have no idea how their ordinary possessions—everyday items such as televisions and computers—actually work. And we live in a time when the issues raised by science, whether in the form of the genetic revolution or electronic eavesdropping, are likely to become even more intertwined with basic decisions about how we live.

Both scientists and attorneys begin with hypotheses. But interaction between the worlds of law and science in the courtrooms makes clear that they represent two very different traditions. Clashes are already common between the truth-seeking world of science and the justice-serving institutions of the law; they are likely to intensify in the future. Each field, perhaps, sees the other as easy prey for the ancient intellectual trap once expressed as "an error is the more dangerous the more truth it contains."

At the heart of the problem, of course, are the rapid technological and scientific advances that make it ever more difficult for those involved in the judicial system and for the citizens who serve on juries to understand the complex information presented by expert witnesses. Cases involving litigation over risk and evidence, such as DNA test results, are ripe with the potential for misunderstanding and confusion. The issues raised by such cases often go beyond legal procedures and the scientific method. Moreover, both legal professionals and scientists are part of dynamic social institutions that are touched by and touch upon other fields; the "facts" by which they are guided seem, at least to lay people, to be changing all the time.

In the pages that follow, Sheila Jasanoff provides a broad and insightful examination of many of the issues raised by judicial activity in this area—an area she describes as "situated at the intersection of law, science, politics, and public policy." It may well be that today much of what most Americans learn about science comes from the coverage of science in our courtrooms. The risks from secondhand smoke and microwave technology are elucidated in each case involving damages from these products; genetic research is a little clearer for us every time there is a case involving paternity; and medical advances are explained in every case involving malpractice or the right to life. That information becomes part of our knowledge base as we choose sides in political controversies involving these issues. It helps inform the debates over environmental issues wherein calculations of risk from new technologies are measured against their effects on our nation's drive for economic and technological growth.

When Jasanoff first came to the Twentieth Century Fund

with this project, she was at the beginning of a career that has helped shape a new field of study. Since then she has established the first major interdisciplinary university department of science and technology studies at Cornell University. In this book she examines the issues that we will face as the world becomes ever more complex and suggests ways for preparing to meet that complexity, such as educating those involved in the legal system or establishing alternatives to litigation. We are grateful to Jasanoff for an informed and provocative tour of these difficult issues.

Richard C. Leone, President
The Twentieth Century Fund
June 1995

Preface

This book is the product of a long and still unfinished journey. My purpose is to explore how two of the most powerful institutions in America, science and the courts, interact with each other in the face of technological innovation and political change. Scholarly interest in this relationship is of relatively recent origin, although it is growing, as evidenced by the proliferation of journals and programs of law, science, and technology in law schools around the country. For the rest of the social sciences, however, what takes place in engagements between the truth-seeking agencies of science and the justice-serving institutions of the law remains a largely unexamined problem. As a result, outside the pages of specialist journals there are as yet no canonical approaches to writing about science and technology in the legal process. If, as I propose here, our society is increasingly defining itself through conflicts that are at once scientific, technological, and legal, then the academic voices that could interpret or give coherence to these multiple, loosely connected acts of self-definition are striking by their absence.

Commentators on law, science, and technology have focused most often on the difficulties faced by judges and juries in recognizing "good science" and "legitimate expertise," both of which are presumed to exist unproblematically in a world that is independent of the day-to-day workings of the law. Such writing takes for granted the capacity of scientists to settle factual disputes and to distinguish legitimate from illegitimate claims without on the whole being swayed by external influences. Science, in this reckoning, is not in itself a questionable source of authority; the problem, if any, lodges in the legal system's inability to recognize the proper emissaries of science and defer to the messages they carry. This view of science as an autonomous and largely self-regulating field of inquiry is often joined to a view of the law that conceives of scientific validity as a precondition for rendering justice. From this perspective, it is easy to reduce the law's obligation in relation to science to a simple, two-step prescription: courts or other legal institutions should first seek out the findings of mainstream science and then incorporate them into their adjudicatory decisions.

The approach I take here assumes to the contrary that scientific claims, especially those that are implicated in legal controversies, are highly contested, contingent on particular localized circumstances, and freighted with buried presumptions about the social world in which they are deployed. The institutional setting of the law shapes the representation of legally relevant scientific claims at many points, beginning with the articulation of standards for what counts as valid science within the legal process. In other words, ideas of truth and ideas of justice are co-constructed in the context of legal proceedings. Legal controversies over "good science" and how to find it accordingly serve for me as starting points for a more discursive inquiry. I wish to understand how the legal process mediates among conflicting knowledge claims, divergent underlying values, and competing views of expertise in a democratic society.

The law plays an equally fundamental role in constructing the fit between technological artifacts and their social context. Advances in the realm that is conventionally labeled "technological" inevitably require the readjustment of existing human behaviors, institutions, and relationships. They enable new modes of conduct—and sometimes foreclose old ones—thereby calling into question notions of fundamental significance to the law, such as agency, causality, rights, responsibility, and blame. The redefinition of concepts such as these around new scientific and technological developments, and the underlying successes and failures of judicial skill and imagination, are among the central concerns of this book.

My account of the law's interactions with science and technology has been influenced by a deepening commitment to the field of science and technology studies over the past ten years. Work in this field calls attention to the negotiable boundaries of many things whose hardness we ordinarily take for granted, such as facts, institutions, social roles, and even inanimate objects. A core project of science and technology studies has been to display the fluidity of the divisions among the social, material, and natural worlds, showing that much of what we know through science or use as technology is produced and given solidity through socially accredited systems of rhetoric and practice. Science, in particular, emerges from this analysis not as an independent, self-regulating producer of truths about the natural world, but as a dynamic social institution, fully engaged with other mechanisms for creating social and epistemological order in modern societies. Within this analytic framework, the interplay of law and science acquires particular significance, since the law's power to articulate social norms becomes tightly interwoven with science's efforts to declare unchanging truths about the nature of our physical world and our own selves. Seen close up, legal disputes around scientific "facts" often appear as sites where society is busily constructing its ideas about what constitutes legitimate knowledge, who is entitled to speak for nature, and how much deference science should command in relation to other modes of knowing.

The questions I grapple with in the following pages are situated at the intersection of law, science, politics, and public policy. The idea of "social construction," which is central to current scholarship in science and technology studies, provides a conceptual connection among chapters dealing with topics as varied as expert witnesses, judicial review, toxic torts, and the regulation of new technologies. In each context, legal proceedings function as a medium for constructing and stabilizing particular orderings of science and technology in society. At the level of legal analysis, I am particularly interested in the formal and informal techniques by which courts legitimate some, and exclude other, possible interpretations of technical expertise, claims, products, and processes. From legal scholarship, as well as from science studies, I draw my concern for the ways in which general claims and principles emerge from the particularities of specific cases and controversies. The book's political dimension derives from the fact that constructions of science and technology in the legal system invariably redraw the lines of power and authority, as when the law opens up new areas of technical decisionmaking to review or control by nonexpert publics. One of the book's major conclusions arises from its mix of disciplinary perspectives: the legal system, I argue, has been instrumental in creating and sustaining public understandings of science and technology in the very processes of "using science" to resolve technical controversies. It follows that one cannot fully comprehend the place of science and technology in American political life without closely attending to their deployment in the legal process.

My thoughts about this book have benefited enormously from conversations with many friends, colleagues, and students in science and technology studies. I am especially grateful to Dorothy Nelkin, whose early and enthusiastic encouragement led me to this field. Michael Dennis, Evelleen Richards, Wesley Shrum, Laurence Trancredi, and Brian Wynne all provided much-needed incentives to carry on with the project when self-doubt and other pressures threatened to overwhelm it. Many colleagues in the realms of law and science contributed with valuable information and still more valuable reality checks. My contacts, both formal and informal, with the American Association for the Advancement of Science and the American Bar Association through the National Conference of Lawyers and Scientists were particularly helpful in expanding my awareness of the issues faced by practicing judges, lawyers, and expert witnesses in technically complex cases. I owe a special debt to Bert Black, Joseph Cecil, and Edward Gerjuoy for their unfailingly thoughtful and informative collegiality, and to Mark Frankel, Deborah Runkle, and Albert Teich for drawing me into many stimulating interactions through AAAS. Thomas Gieryn, Michael Reich, Joseph Sanders, Peter Schuck, and several anonymous reviewers provided exceptionally perceptive and generous readings of the penultimate draft of the manuscript.

Work on the book was supported by the Twentieth Century

Fund at a time when external recognition was especially important to a struggling academic in a then little-known interdisciplinary research program. I am deeply grateful for this support and for the continued, near-saintly patience of Beverly Goldberg and her colleagues Pamela Gilfond and Roger Kimball, through the protracted recasting and revision that followed the award. My assistant Deborah Van Galder was as always a source of efficiency and strength, as were my research assistants Nora Demleitner at Yale Law School and Lauren Gelman, Jennifer Huang, Olive Lee, Tania Simoncelli, and Robert Speel at Cornell. I am also indebted to Robert Gulack for thoughtful critical comments on an early draft. Most of all, I thank the members of my family who bore with the book and its author through many tribulations—my mother, Kamala Sen; my husband, Jay; my son, Alan; and, not least, my daughter, Maya, whose life has unfolded in tandem with my life in the law.

Although the book has been enriched by the help and insights of many people, choices of content and interpretive strategy have been mine alone. Thus, I have not attempted to be comprehensive in my descriptions of contemporary legal encounters with science and technology; many important topics, such as intellectual property law or psychiatric, medical, and social science evidence, receive only passing attention in the following pages. Instead, this book represents an initial attempt to open up what I hope will be a continuing and mutually enriching conversation between science and technology studies and the law. I see it not as a final resting place but as a temporary vantage point from which to survey the surrounding terrain more clearly and from which others may be inspired to climb further and reveal new heights. The book will serve its purpose very well if it makes even a few of its readers look with reawakened interest at the changing configurations of science, technology, and the legal order in our vibrantly litigious society.

1

The Intersections of Science and Law

American political culture derives its distinctive flavor as much from faith in scientific and technological progress as from a commitment—some might even say an addiction—to resolving social conflicts through law. These powerful cultural predilections have brought the institutions of science and technology into turbulent confrontations with the legal system, raising doubts whether the nation's intense concern for health, safety, environmental protection, and traditional moral values can be reconciled with its yearning for sustained economic growth and an endless technological frontier. Disenchanted with legal approaches to problem-solving, influential critics in universities, industry, and government have called upon the law to adopt a more deferential or hands-off attitude toward science and technology. The legal system is again being urged to leave the resolution of scientific disputes to scientists, and even the much-maligned idea of a "science court" has enjoyed a minor renaissance.

How should policymakers—lawyers, politicians, scientists, and informed citizens alike—come to terms with this groundswell of discontent? Does the fault, if any, lie exclusively with the courts in the often controversial encounters between law and science? Or do our own expectations concerning science and technology need to be modulated in the light of what we know about the limits of adjudication and the nature

1

of scientific inquiry? I address these questions by looking more closely at what actually takes place in legal disputes involving science and technology. How do the distinctive actors, institutions, procedures, and formal languages of the law shape the meanings that science and technology acquire in people's everyday lives? How are public understandings of science and technology altered or, alternatively, exploited when legal resources are brought to bear on the resolution of technical controversies? And how does litigation constrain or redefine our collective ability to control the development of technology? In exploring a variety of techno-scientific disputes in the legal arena, I will reexamine many of the received opinions underlying scholarly writing on law and science and will consider the benefits of a more reflective, self-aware response to science and technology by the legal system.

America's preoccupation with progress through science and technology appears, at one level, to be solidly grounded in historical achievements. A century of inventions has enlarged our capabilities and improved our quality of life in myriad unpredictable ways. At every turn we encounter new material indicators of progress: air bags and antilock brakes, electronic mail, fax machines and bank cards, heart transplants and laser surgery, genetic screening, *in vitro* fertilization, and a burgeoning pharmacopeia for treating mental and physical illness. In just one generation the space program has expanded the physical frontiers of human experience, while discoveries in the biological sciences have revolutionized our ability to manipulate the basic processes of life so as to fight infertility, aging, hunger, and disease. Scientific research continues to attract generous public funding, even in a time of political skepticism and budget deficits. Although some "big science" projects, such as the Strategic Defense Initiative (SDI, or "Star Wars") and the superconducting supercollider, lost their appeal with the end of the Cold War, others, such as the space station and the project to map the human genome, continued to attract governmental support despite vigorous criticism from the political and scientific communities.

Increasing knowledge, however, has also reinforced some archetypal fears about science and technology that overshadow the promises of healing, regeneration, material well-being, and unbroken progress.[1] Rachel Carson's *Silent Spring*[2] achieved global bestsellerdom in the 1960s with warnings of a future in which animals sicken, vegetation withers, and, as in Keats's desolate landscape, no birds sing. Genetic engineering, rightly seen as one of the great scientific breakthroughs of this age, has been etched on the public consciousness as the technique by which some modern Dr. Frankenstein may fatally tamper with the balance of nature or destroy forever the meaning of human dignity. Communication technologies speed up the process of globalization but threaten to dissolve the fragile ties that bind individuals to their local communities. The mushroom cloud, nuclear winter, the "hole" in the stratospheric ozone layer, the "greenhouse effect"—these disturbing images all suggest that our civilization's Faustian thirst for knowledge has outstripped our capacity to foresee and ward off the effects of dominating nature too completely.

Opinion polls and the popular media reflect the duality of public expectations concerning science and technology. A 1992 Harris poll showed that 50 percent or more of Americans considered only science and medicine to be occupations of "very great prestige," but that these ratings had fallen by 9 and 11 percentage points, respectively, since 1977.[3] According to another survey, the proportion of people expressing a great deal of confidence in scientific institutions remained fairly steady, at 36–45 percent, between 1973 and 1993. Yet in 1993 more than 45 percent of the public felt that there would be a nuclear power plant accident and significant environmental deterioration in the next twenty-five years; slightly smaller percentages expected to see a cure for cancer and a rise in average life expectancy in the same period.[4] Ambivalence prevailed in Europe as well. In polls conducted in 1989, more than 75 percent of respondents believed that science improves the quality of life and should be supported by the state, but 28–65 percent disagreed or disagreed strongly that scientists could be trusted to make the right decisions.[5] Following the unprecedented success of the film *Jurassic Park* in 1993, a science reporter for the *New York Times* observed that, at least in the eyes of the popular media, "drug companies, geneticists and other medical scientists—wonder-workers of yesteryear—[were] now the villains."[6]

Much of the recent entanglement of law and science reflects the American public's determination to bring under control the darker side of technological mastery: risks to public health and the environment, to individual autonomy and privacy, and to community and moral values. From Tocqueville to the present, commentators on American culture have called attention to this country's particular penchant for resolving political controversies and achieving social order through law. It is hardly surprising that in an age of anxiety about the products of science and technology the U.S. public has increasingly turned to law to reassert control over the processes of scientific and technological change or to seek recompense for the failed promises of technology.

Yet, as we near the end of the century, law no less than science has lost much of its progressive mystique. Litigation today is perceived more as a cause of than a cure for some profound malaises of American life. The soaring costs of legal proceedings, combined with attacks by political conservatives on judicial activism, have reduced the credibility of the courts; an excess of law is blamed for many of the problems that have beset U.S. science and technology in the recent past, from plummeting medical school applications in the 1980s (a trend since reversed) to corporate bankruptcies, delays in product introductions, and lagging industrial competitiveness. Prominent social critics, including not only scientists but also members of the bench and bar, argue that the power of the courts must be checked if the United States is to be pulled out of a downward economic spiral.[7] In their tactically brilliant, ten-point "Contract with America" of September 1994, the Republican Party promised a Common Sense Legal Reform Act that would introduce "loser pays" laws, limit punitive damages, and reform products liability laws so as to discourage litigation. The erosion of faith in legal processes and institutions reflected in these developments is as much a part of the context for this book as is the public's often expressed distrust of technical experts and their undemocratic authority.

Truth or Justice?

Complaints about the legal system's handling of problems related to science and technology correspond in general terms to two distinct yet well-established traditions of science policy analysis: one concerned with "science in policy" and the other with "policy for science."[8] While one can question the basis for this binary distinction, it continues to dominate public perceptions about the mismatch between the needs of science and technology and those of the legal process.

The project of "science in policy" has encompassed repeated proposals to "improve" the use of science in legal decisionmaking by reforming the selection of expert witnesses, reeducating judges and juries, and changing the standards for validating technical evidence. Critics often note that American judges, juries, and lawyers know on average very little about the social organization and processes of science, still less about basic scientific concepts such as "statistical significance," and almost nothing about the substantive content of particular scientific fields. Yet these "technically illiterate" fact-finders, who understand neither the substance nor the methods of science, are increasingly called upon to discriminate among sophisticated technical arguments. Lacking adequately trained gatekeepers, the legal system allows itself, in the view of some critics, to be swamped by "junk science," while truth and rationality fall victim to the manipulative dynamics of the adversary process.[9] Cross-examination and the legal rules of evidence operate only as recipes for obfuscation, while the control of expert witnesses by litigants brings to court only the most extreme and unrepresentative opinions about the technical issues at stake in litigation.

Daniel Koshland, an editor at *Science*, one of the nation's premier journals for the biomedical sciences, occasionally used his position to direct satirical barbs against the legal system. In the following extract, a naive "Science" learns the facts of legal life from "Dr. Noitall":

> *Science.* So the judicial system is not a system to get at the truth as simply as possible.

Noitall. Finally you understand. The judicial system is an adversarial system in which clever lawyers match wits with one another. If a lawyer defending a mobster murderer can show a technical discrepancy that gets his client free, the lawyer is widely admired even though a killer has been freed.[10]

Though written by a scientist for the consumption of other scientists, such storytelling nevertheless exemplifies a significant form of boundary maintenance between science and the law. On the pages of *Science*, "science" emerges as unswervingly committed to the truth, while the law is shown as intent on winning adversarial games at any cost.

The critical project concerned with "policies for science" has focused mainly on the inefficiencies of judicial decisionmaking as an instrument for managing technology. Observers in this camp question whether the judiciary is institutionally capable or constitutionally empowered to make policy on issues such as biotechnology, nuclear power, or new medical and reproductive technologies. Because of jurisdictional constraints, neither the state courts nor the lower federal courts seem qualified to develop policy on a national scale. Moreover, the retrospective and case-by-case methods of adjudication seem to many to be fundamentally incompatible with the nation's need for forward-looking responses to science and technology. The impact of the courts on innovation, liability, the selection and funding of substantive research programs, and the regulation of technological risk appears from this standpoint to be a significant obstacle to progress.

Both of these standard framings converge in their assumption that science and, to a lesser extent, technology possess an inner logic, an autonomous framework of validation and control, that operates irrespective of the law and does not need to be subjected to the law's normative concerns or institutional practices. Prescriptions for injecting "good science" into legal decisions—for example, by delegating scientific issues to scientists or making legal actors more technically literate—are a manifestation of our society's deep commitment to a rational, and hence reliably objective, policy process. Those who would lower the legal barriers to scientific creativity and technological innovation are similarly wedded to the idea that technology policy is best handled by expert, rational decisionmakers. Each of these conceptualizations attributes, in my view, an untenable firmness to the boundary between science and technology on the one hand and law and policy on the other. Recommendations for institutional separation, such as the formation of science courts, along with demands for better science in the legal process, chronically overestimate the power of experts to rationalize moral and political choices about science and technology.

The Cultures of Legal and Scientific Inquiry

The contrasts between law and science are often described in binary terms: science seeks truth, while the law does justice; science is descriptive, but law is prescriptive; science emphasizes progress, whereas the law emphasizes process.[11] These simplified characterizations restate in varying ways the insight that fact-finding in the law is always contingent on a particular vision of (and mechanism for) delivering social justice. Scientific claims, by contrast, are thought to lack such contingency. Although its conclusions may be speculative, provisional, and subject to modification, science is ordinarily seen as set apart from all other social activities by virtue of its institutionalized procedures for overcoming particularity and context dependence and its capacity for generating claims of universal validity. Not surprisingly, then, comparisons between science and the law often celebrate science's unique commitment to systematic testing of observations and its willingness to submit its conclusions to critical probing and falsification.[12]

The representation of law and science as fundamentally different enterprises has given rise to two strikingly recurrent themes in legal writing about science: that of the "culture clash" between lawyers (or legally trained bureaucrats) and scientists, and the corollary that the culture of law should strive as far as possible to assimilate itself to the culture of science when dealing with scientific issues. Cultural disparities are offered to explain the discomfort that scientists feel when asked to express their technical judgments in the heavy-handed categories used by regulators, or when they are cross-examined by lawyers who dredge up minor inconsistencies to impugn the experts' credibility.[13] Analysts of the law, it appears, feel almost as uncomfortable about these confrontations as the scientists they describe. Prescriptions for overcoming the alleged cultural conflicts between law and science range from injunctions to the courts to borrow the standards scientists themselves use in distinguishing "real science from . . . pale imitations"[14] to suggestions for professional mediators, such as science bureaucrats or science counselors, who will enable each culture better to understand the other.[15]

Missing from this literature is any but the most cursory attention to the commitments and practices that actually constitute the culture of science. Still less effort has been dedicated to understanding the scientific subcultures that have coalesced in and around the processes of adjudication. In the following chapters I will use insights from science and technology studies, supplemented by case studies of particular areas of legal development, to challenge the abstract and idealized view of science expounded by most proponents of the "culture clash." Taking real decisions and controversies rather than wishful scenarios as the point of departure, I will argue that the cultures of law and science are in fact mutually constitutive in ways that have previously escaped systematic analysis. Understanding how these institutions *jointly* produce our social and scientific knowledge, and our relationships with technological objects, is indispensable to any effective attempt at policy reform.

As formal systems of inquiry, law and science have several important features in common. Each tradition claims an authoritative capacity to sift evidence and derive rational and persuasive conclusions from it. The reliability of observers (or witnesses) and the credibility of their observations are of critical concern to both legal and scientific decisionmaking. Unlike organized religion, neither science nor law owes allegiance to a single dogmatic authority. In both fields, rules governing the assessment of facts occasionally undergo massive shifts—in science through the work of paradigm-transforming pioneers[16] and in the law (ordinarily but not always) through the actions of legislatures. Normal progress within each discipline occurs through a decentralized, silent revolution brought about by individuals making decisions at the frontiers of established doctrine in accordance with their personal understanding of the existing tradition.[17]

The considerable differences between scientific and legal thinking are most apparent in their approaches to fact-finding. Science, as conventionally understood, is primarily concerned with getting the facts "right"—at least to the extent permitted by the existing research paradigm or tradition. The law also seeks to establish facts correctly, but only as an adjunct to its transcendent objective of settling disputes fairly and efficiently. This basic dichotomy accounts for a number of secondary contrasts. Because the law needs closure, the process of legal fact-finding is always bounded in time: inquiry has to stop when the evidence is exhausted. The judicial inquirer cannot postpone a decision by choosing to wait for more evidence. As John Ziman, British physicist and sociologist of science, has noted, "If we are forced to a premature opinion on a scientific question, we are bound to give the Scottish verdict *Not Proven*, or say that the jury have disagreed, and a new trial is needed."[18] The law, by contrast, must take a position based on the facts at hand, however premature such a decision may appear in the eyes of scientists.

Fact-finding in law proceeds through a form of ritualized courtroom discourse that subjects the scientist's firsthand reporting of observation and experiment to additional conceptual and rhetorical filters. What the legal fact-finder "knows" is a function of what the witnesses in a proceeding choose to relate in court in answer to questions posed by lawyers. British mystery writer R. Austin Freeman wryly commented on this highly restricted form of knowing in a 1911 novel: "The scientific outlook is radically different from the legal. The man of science relies on his own knowledge and observation and judgment, and disregards testimony . . . A court of law must decide according to the evidence which is before it; and that evidence is of the nature of sworn testimony. If a witness is prepared to swear that black is white and no evidence to the contrary is offered, the evidence before the Court is that black is white, and the Court must decide accordingly."[19] Freeman satirized, but around a kernel of truth. "Science," for the law's purposes, is simply the composite of testimony presented in and around an adjudicatory proceeding, and its quality depends heavily on the skill and intentions of the lawyers who elicit the presentation. The facts that the law constructs (or reconstructs) are thus necessarily different from the facts that scientists construct to persuade their peers in their own rhetorically and procedurally distinctive surroundings.

To serve its need for decisive endings, the law has devised a complex system of rules and practices for choosing what to believe when facts are uncertain; these rules and practices by definition are not "scientific." They include, to start with, the rules by which the legal system determines what evidence and which witnesses are relevant to the dispute at hand.[20] Another body of legal rules addresses the problem of making decisions on the basis of conflicting evidence. For example, in civil cases the legal system places the "burden of proof" on the plaintiff. In order to prevail the plaintiff must prove his claim by a "preponderance of the evidence"—in other words, more than 50 percent of the evidence must be in the plaintiff's favor. This requirement is a way of ensuring that even in those borderline cases in which the evidence is perfectly balanced the legal factfinder will have an orderly basis for deciding between the disputants. Science under the same circumstances would be neither willing nor able to declare a winner. Administrative decisions generally call for a lower standard of proof, whereas criminal trials demand something closer to scientific certainty ("beyond reasonable doubt"). A contrafactual or contrascientific conclusion can, in appropriate circumstances, be declared the "right" conclusion from the standpoint of the law. In criminal proceedings, for instance, evidence deemed highly relevant to a scientifically "correct" determination of guilt or innocence might be excluded in order to protect individuals against coercion by the state, thereby producing a technically "incorrect" but morally just outcome.

Even in civil cases, the legal system's allegiance to values other than those of science may open the way to decisions that look like sheer irrationality. For example, in a 1946 paternity case against Charlie Chaplin, a jury held the actor liable, and the court ordered him to pay child support, even though blood-group evidence showed he could not have been the father.[21] Francisco Ayala, a distinguished biologist, and Bert Black, an engineer-lawyer, cite this as a bizarre decision that flies in the face of scientific knowledge.[22] Michael Saks, a social psychologist and expert on evidence, takes a more moderate stance, pointing out that "the jury may have doubted the manner in which the blood test was carried out, the underlying science, or the honesty of the expert witnesses."[23] Equally, Saks notes, the jury's sense of justice could have affected the outcome. Biological relationships, after all, do not always control in determining an adult's financial or custodial responsibility for a child. To ensure support for children, state laws have traditionally presumed that a child born to a married woman is her husband's legitimate offspring. Perhaps the jury analogized Chaplin's position to that of the canonical husband, taking into account his vastly superior economic standing in comparison with the child's mother. Under any of these readings, the jury's refusal to accept the scientific denial of Chaplin's paternity could properly be characterized as social wisdom rather than scientific illiteracy.

Adjudication and Technology Assessment

The emphasis on legal fact-finding in recent discussions of law and science has tended to obscure the role of courts in managing technology, although in the long run this role is arguably more important to the evolution of modern industrial democracies. Scholarly writing on the intersection of our legal and technological cultures has ascribed to courts a largely reactive role; that is, courts are seen mainly as instruments for remedying the negative impacts of technology. Conventional institutional analysis supports this view. Courts, after all, cannot initiate action on their own but must await complaints from aggrieved parties. Unlike legislation, adjudication proceeds case by case and retrospectively—that is, after harm has occurred. The timing of litigation thus precludes courts in theory from the primary goal of technology assessment: "rational choice at the earliest possible stage of events."[24] Yet, as we shall see, U.S. courts are often the first social institutions to give public voice and meaning to formerly inaudible struggles between human communities and their technological creations. Technological trajectories are therefore importantly steered by events within the legal system.

Legal thinking about technology and the courts has been informed for the most part by an unexamined technological determinism. Thus, in a recent study of law and culture even so perceptive an interpreter of the American legal order as Lawrence Friedman accords little attention (only 10 pages in 206 pages of text) to the relationship between technological and legal change, and he describes the impacts of technology only in the conventionally deterministic language of cause and effect.[25] The effects include, in his telling, a heightened pace of change, an explosion of legal forms and structures, and an increase in expressive individualism, manifested in American society's growing sense of legal entitlement.

Others who also see the law as concerned primarily with technology's impacts have charged that the legal process allows its concern for individual justice to override larger societal interests, such as the industrywide or cross-sectoral effects of new liability rules. Peter Huber, one of the most outspoken critics of U.S. tort law, argues that the regulatory force of litigation systematically and disproportionately penalizes "public risks"[26] and gives preference to larger "old" risks over smaller "new" ones.[27] Huber includes in the category of public risks large power plants, mass-produced vaccines, and jumbo jets— all technologies which in his view have lowered the risks of disease and death from older, "cottage industry" alternatives such as wood stoves, automobiles, and exposure to naturally occurring toxins. Huber believes that adjudicatory decisions consistently disfavor threats to health and safety that are centralized, mass-produced, and lie outside the risk bearer's individual control. By overcompensating the victims of new technologies and discouraging their proliferation, courts increase the total probability of harm to humans and the environment.

Huber evidently presumes that people prefer those technologies that most reduce the number of accidents and that courts should simply respect these preferences. But study after study of risk perception has pointed out, to the contrary, that the public is often more concerned about the social and cultural dimensions of risk than about actual or predicted numerical impacts. Thus, people will tolerate a higher probability of death or injury from activities that they feel they can meaningfully control (smoking, eating, automobile driving) than from activities that heighten their sense of powerlessness or distrust (nuclear power, pesticide use, air transportation).[28] Doing justice in a democratic society requires courts to respect such entrenched normative positions even if they cut against the managerial preferences of the nation's scientific and technological elite.

Courts, however, bring distinctive institutional limitations as well as competencies to their construction of the relationship between material objects and social needs, and both characteristics should be addressed in a fair evaluation of their performance. Products liability litigation may produce socially insupportable dislocations when a single "bellwether" decision exerts effects far beyond the factual content in which it arises. Claims against vaccine manufacturers in the 1980s created an uncertain legal environment in which drug companies considered it more prudent to cease production than face the prospect of huge damage claims by a small number of plaintiffs.[29] The insurance industry's reluctance to assume responsibility for extraordinary damages magnified the dampening effect of tort litigation, in some cases beyond acceptable limits. In 1986, for example, several thousand patients suffering from a rare neuromuscular disorder were deprived for many months of a beneficial experimental drug because its producer could not obtain products liability coverage.[30] Congress in the same year enacted a vaccine compensation bill to transfer such claims out of the courts and put a cap on awards for pain, suffering, or death;[31] the Republican "Contract" mentioned earlier promised to generalize damage caps to areas other than vaccine compensation.

The political legitimacy of courts is equally vulnerable to

challenge when judges have to second-guess expert administrative agencies. In the early years of biotechnology development, for example, environmental activists successfully delayed the commercialization of some genetically engineered
products through legal action. By deciding in favor of the environmentalists in a few such cases, courts legitimated public
opposition to the executive branch's policy of promoting a
technology deemed essential for the nation's competitiveness.
Although public interest groups applauded these initiatives,
others questioned the validity of judicial intervention in such a
specialized and economically vital area of science and technology policy. Their argument (critically appraised in Chapter 7)
rested in part on a narrow interpretation of the courts'
policymaking function and in part on the view that courts
threatened science by giving equal time to scientifically
groundless fears and objections.

The notorious cost and inefficiency of the judicial process
also prompt concern about the role of courts in shaping technology policy. Legal redress for technological accidents is neither cheap nor speedy. In complex cases, the expense of producing evidence and hiring expert witnesses adds significantly
to the already heavy burden of attorneys' fees and court costs.
A lawsuit involving claims of pollution-related disease can easily consume millions of dollars and numbers of years, especially if it proceeds to trial. One of the longest jury trials in the
nation's history arose from a suit against Monsanto by residents of Missouri who claimed to have been injured by dioxin.[32] Three years of inconclusive testimony in that case convinced many participants that the legal system was helpless
against abuses by determined and well-financed litigants. In
1990 the nation's longest-running criminal trial, a California
child molestation case, ended without a conviction when jurors decided that they could not extract a clear interpretation
of the facts from 60,000 pages of testimony by 124 witnesses.[33]

Financing complex litigation, moreover, demands ad hoc arrangements that create problems of inequity and unequal access. In a toxic tort case in Woburn, Massachusetts, the citizen
plaintiffs were able to pursue their claims against one defendant, W. R. Grace, mainly because another firm had pre-

viously settled out of court for $1 million.[34] Peter Schuck's illuminating account of the massive lawsuit brought by Vietnam veterans against manufacturers of Agent Orange shows that the investments made by the lawyers, in time and money, completely overshadowed the ill-defined, uncertain, and circumscribed claims of individual plaintiffs. The Agent Orange litigation slowly evolved into what Schuck terms "a lawyers' case," with the veterans' interest in compensation and retribution taking second place to the lawyers' financial and professional interests; a similar issue seemed sure to present itself in a massive class-action lawsuit filed by a consortium of close to sixty law firms against the tobacco industry.[35] The tort system, as well, is one of the world's most costly and inefficient mechanisms for returning compensation to plaintiffs. A well-known study of asbestos lawsuits by the Rand Institute of Civil Justice suggested that toxic tort liability entailed exceptionally high transaction costs. From the early 1970s to 1982, every $1 paid to asbestos plaintiffs as compensation entailed an additional $1.71 in litigation expenses by plaintiffs, defendants, and their insurers.[36] In other words, only 37 percent of the total expenditures were recovered by the victims. Other studies have further substantiated this finding.[37]

Science and Technology in Court

To probe more deeply the engagements between science, technology, and the legal system, we need a working map of present-day litigation patterns in this area. Liability actions remain central to any map, since no serious technological mishap can occur in the United States without triggering a legal response.[38] Products liability litigation is by now so commonplace that it scarcely attracts attention unless the plaintiffs are particularly vulnerable or numerous, the defendant's conduct is especially egregious, or the monetary stakes are exceptionally large. Yet the very pervasiveness of products liability creates room for contradictory moral and political interpretations. For some, the liability action is the only device by which impersonal, profit-mad corporations can be held accountable to their exploited and economically disadvantaged victims. In this spirit, Marc Galanter, an authority on the American tort system, celebrates its evolution through the twentieth century into a mechanism for delivering high accountability as well as high remedies to injured citizens.[39] Cases that support his reading include in recent years suits against Ford Motor Company and General Motors by victims of gas tank explosions, against Manville Corporation by asbestos workers, and against A. H. Robbins by women injured by the Dalkon Shield. Others, more critical, see products liability simply as a convenient causal framework within which people can rationalize their often unspeakable misfortunes by building tenuous connections to improbable technological causes. Claims that have attracted this charge include lawsuits by alleged victims of exposure to dioxin, breast implants, and electromagnetic fields (EMFs) from power lines. Widely publicized because of their complexity and tragic origins, these cases reinforce the image of the United States as a forum where scientific standards of proof are routinely flouted and where irresponsible juries destroy productive industries through ignorant, ill-considered verdicts.

But courtroom activity surrounding science and technology now encompasses a web of social adjustments extending far beyond liability. The rapidly growing roster of technology-related disputes challenges any simple characterization of the role courts play in mediating the fit between science, technology, and society. As the following examples suggest, the law today not only interprets the social impacts of science and technology but also constructs the very environment in which science and technology come to have meaning, utility, and force. These cases begin to alert us to the subtle connections that exist between conflicts over knowledge, the traditional preserve of science, and conflicts over responsibility, the classic preserve of the judiciary.

• A 1993 decision by New York State's highest court held that a claimant could seek damages for a drop in property value caused by public fear of a right-of-way for a high-voltage power line. The claimant was not required to prove that there were medically or scientifically reasonable grounds for the phobia concerning the effects of electromagnetic fields. The court felt that the economic question of the loss in market value could be resolved without being "magnified and escalated by a whole new battery of electromagnetic power engineers, scientists or medical experts."[40]

• In 1993 the U.S. Supreme Court overturned a seventy-year-old federal rule governing the admissibility of expert testimony and announced, in *Daubert v. Merrell Dow Pharmaceuticals, Inc.*,[41] new criteria by which judges ought to distinguish between valid and invalid scientific evidence. Among the many issues affected by *Daubert* was a line of cases challenging the admissibility of the so-called horizontal gaze nystagmus test administered by police officers around the country to drivers suspected of intoxication. The objective of the test is to correlate a visual inspection of the suspect's eyeball movements with a measure of intoxication. A California court allowed the testimony but observed that the officer offering it was not a scientist because he "drew his generalization from experience, not from experimentation, and did not attempt to quantify the relationship he observed."[42]

• A 1994 article by two prominent scientists in the prestigious British journal *Nature* declared that all scientific controversy over the technique known as "DNA fingerprinting" could henceforth be laid to rest.[43] The attempt by two formerly battling experts to create a consensus around this important forensic technique was widely seen as a response to the notorious murder trial of O. J. Simpson, the football hero and media personality who had recently been charged with killing his ex-wife Nicole Brown Simpson and her friend Ronald Goldman.

• A federal district court held in March 1993 that President Clinton's Task Force on National Health Care Reform should be viewed as an advisory committee subject to the open meetings requirements of the Federal Advisory Committee Act. Although most of the task force members were federal employees, it was chaired by First Lady Hillary Rodham Clinton, who was not an officer or employee of the federal government. In support of its decision, the court cited the public's right to know what information was being given to the task force, by whom, and at what cost.[44]

• Two scientists working for the National Institutes of Health developed a computerized "plagiarism detector" to measure the degree of overlap among texts in lawsuits over copyright violations. Their zeal to perfect such a technique reflected the legal system's increasing involvement with scientific misconduct in the late 1980s.[45]

• The California Supreme Court in 1993 denied a plea by a dead man's two adult children to order the destruction of sperm that he had bequeathed to the woman he had lived with before he committed suicide. Reflecting the same changing social norms about reproduction, parenthood, and family, a Florida woman who had been married only two weeks at the time her husband died in 1994 announced that she hoped to start a family using sperm collected shortly after his death. Also in 1994, a gay man was granted visitation rights to a child he had "fathered" by donating sperm to a lesbian couple. And in Louisiana, a child born to a widow through artificial insemination a year after her husband's death was declared illegitimate.[46]

• Identification of the AIDS (acquired immune deficiency syndrome) virus, the development of a blood test, attempts to characterize and contain the risk of infection, and efforts to manufacture a vaccine produced unprecedented legal controversies. In Needham, Massachusetts, an AIDS victim employed by the New England Telephone Company filed a $1.75 million lawsuit charging that the company had discriminated against him on the basis of a handicap, breached his privacy, and coerced him not to return to work.[47] A settlement permitting the plaintiff to return to work led to a protest walkout by fellow technicians who remained convinced that his presence posed serious risks to their safety. On the international scene, France's prestigious Pasteur Institute sued the U.S. National Institutes of Health (NIH) to establish credit for the discovery of the AIDS virus and to claim a share in royalties from patented blood tests for the disease.[48] The controversy temporarily subsided when the French and American heads of state agreed that scientific credit should be shared, but it resurfaced as U.S. investigators sought to determine whether the NIH "co-discoverers" had been guilty of misconduct.[49]

• Developing technologies for extending or preserving the life of the human fetus gave rise to a bizarre series of lawsuits. A judge in Connecticut ordered that the fetus of a woman who had been comatose for nearly three months should not be aborted, because there was "insufficient evidence" to justify such action.[50] In San Diego, California, criminal charges were brought against a woman who had allegedly failed to take proper medical care of her unborn child, in contravention of her doctor's recommendations.[51] According to a hospital autopsy report, the infant died as a result of "fetal distress syndrome caused by maternal drug abuse." Although the charges were eventually dropped, the case achieved notoriety as the first attempt to use criminal sanctions for a woman's treatment of a fetus during pregnancy. And in suburban Virginia, a federal court ordered a hospital to continue providing life-sustaining treatment to a baby born with most of her brain missing. At the time of the decision, Baby K had already been kept alive for many months longer than most anencephalic babies because of her mother's insistence that she be offered mechanical breathing support during respiratory crises.[52]

Many of these cases can be interpreted as struggles over the authority of knowledge. Whose knowledge should count as valid science, according to what criteria, and as applied by whom? When should lay understandings of phenomena take precedence over expert claims to superior knowledge? Should experts' views about risk and cost structure social relationships (as in the AIDS cases), or should they give way to countervailing nonexpert values (as in the EMF and Baby K cases)? These cases also underscore the extent to which the very production of scientific knowledge and techniques is bound up with developments in the law, from police officers' tests for drunk driving to specialist scientists' efforts to define plagiarism or DNA fingerprinting for forensic purposes. More generally, this assortment of cases also reflects the unpredictable ways in which scientific and technological developments slice into settled social relationships and compel a redefinition, through law, of established rights and duties. In addressing these issues, courts in effect are enlisted into an interactive process of social and technological change; they become partners in society's search for new rules to interpret and restructure an altered array of potentialities.

Guiding Concerns

Accepting the inevitability of judicial rulemaking is not equivalent to approving uncritically the whole of the law's current edifice for resolving disputes with a high technical content. Recognition that science is socially constructed in courtroom settings does not obviate the need for judges to decide in specific cases whose evidence should be admitted or how it should be weighted by the jury. More generally, persistent critical commentary on the courts compels us to reconsider to what extent the social adjustments around science and technology should appropriately be delegated to judicial resolution.

As our understanding of law-science interactions becomes more complex, so necessarily does the task of normative analysis. Prescriptive solutions aimed at maintaining the status quo or at enlarging the dominion of "good science" are difficult to support if one adopts the more dynamic and constructivist analysis of law and science that I am proposing. If we accept the notion that law and science are involved in constructing each other in our society, then where can we turn for criteria that will allow us to assess and improve the interactions between these institutions?

A promising approach is to turn to the courts themselves and look again at the functions that they, as distinctive institutional actors, can best perform in a democratic society of increasing technological complexity. One important task is the *deconstruction* of expert authority. Litigation, as we shall see, is an especially potent resource for making transparent the values, biases, and social assumptions that are embedded in many expert claims about physical and natural phenomena. Exposing these underlying subjective preconceptions is fully as important in a justice system as "getting the facts right." As we review a range of judicial decisions, we will therefore consider how successfully they reveal, and hence empower social criticism of, the possibly unconscious biases of expert witnesses. Courts may play a usefully deconstructive role in relation to technology, as well, by exposing "interpretive flexibility" in the meanings that technology has for different social actors,[53] and by pointing toward new possibilities for social control. It will also be important to ask whether and to what extent the judiciary's own biases concerning science and technology enter into the adjudicatory process, and whether these in turn are available for deconstruction. More generally, we will assess whether litigation is a reliable and effective procedure for disentangling the ethical and social concerns raised by advances in science and technology.

A second and related function that may reasonably be expected from the courts is *civic education* about science and technology. How effectively do courts inform litigants and other citizens, the legal community, and various governmental and nongovernmental institutions about the epistemological, social, and moral dilemmas accompanying technological change? In this connection, it will be especially important to ask how legal fact-finders think about scientific uncertainty and controversy, and how effectively they rationalize their own decisions in the face of conflicts in the evidence. We will also be concerned with the relationship between the judiciary's analysis of science and the overall clarity and consistency of judicial rulemaking. In a similar vein, we will consider the appropriateness of legal framings chosen in relation to particular developments in science and technology (for example, the choice between a strict or a flexible construction of a statute or the choice between preserving or overruling a constitutional standard).

Third and by no means least significant is the criterion of *effectiveness*. When citizens seek redress from the courts, they expect a decision exhibiting certain practical as well as moral attributes. American litigants, in particular, are widely known to bring some quite general expectations to the adjudicatory process. Justice should not only be done but be seen to be done. Litigants should feel they actively had their day in court. Decisions should compensate people for pain, psychic distress, and outrage as well as for pecuniary loss. Compensation should not be indefinitely postponed. An effective legal system has to address these demands on a fairly regular basis. One should also ask, under the heading of effectiveness, how judicial decisionmaking combats threats to individual liberty and security, as well as how litigation influences broader national goals such as innovation and competitiveness.

The methods I use in approaching these issues are based to some extent on traditional legal analysis. Leading cases are reviewed in each chapter both because they represent innovative moments in the law and because they exert the widest influence on public understanding and behavior. But many of the questions addressed in this book, such as judicial boundary-drawing and ideas of expertise, are better illuminated by minor decisions clustering around particular areas of controversy, for example, genetic engineering or toxic torts. Most of the following chapters are therefore organized around specific substantive topics and may be seen in a sense as extended case studies; each one combines the analysis of case law with insights from the secondary literature on the social and political context of litigation. This approach permits a more historical as well as comparative examination of the strategies that the legal system uses in constructing science and technology. Indeed, much of the analysis in the following chapters is not particularly time-bound, although wherever possible I have included striking examples from contemporary legal decision-making. The book, finally, is about judicial styles of reasoning and thought as they both influence and are influenced by science and technology; it does not pretend to provide the last word on any particular area of substantive law.

Chapters 2–5 focus mainly on science and Chapters 6–9 mainly on technology, although the dividing line between the two areas is not absolutely clear-cut. The construction of science within the legal system, both as a social institution and as a system of accredited beliefs, forms the subject of the next four chapters. Here, the major themes are the law's constructive as well as deconstructive treatment of scientific authority and credibility. Chapter 2 surveys the evolution of judicial attitudes in three areas of substantive law: products liability, medical malpractice, and environmental law. Historically, decisions in each of these areas responded to changing public expectations concerning the safety and accountability of technological enterprises; in the process, the courts struck new balances between lay and expert understandings of risk, benefit, safety, and harm. This progression must be understood and critically evaluated as a prelude to discussing more recent controversies involving science and technology.

Chapters 3–5 trace the influence of judicial decisionmaking on the definitions of good science, legitimate expertise, and technical rationality. Informed by perspectives from the sociology of science, Chapter 3 focuses on the recurrent epistemological and institutional conflicts around the use of expert witnesses in the courtroom and the methods that the law has devised for dealing with these problems. The chapter illuminates the multiple contingencies, both cognitive and social, that shape scientific fact-finding in the legal system. Chapter 4 looks at the impact of the courts on the federal regulation of health and environmental hazards, with particular attention to the law's role in making governmental agencies accountable to the public for their technical decisions. Chapter 5 further develops the theme of accountability, this time with regard to relations between scientists and the public. Specifically, the chapter surveys the actions of courts in conflicts arising from scientific and medical research and the confrontation between scientific and religious values.

Chapters 6–9 are concerned primarily with the capacity of the courts to sort out controversies concerning the ethical, social, and cultural implications of technological change. Chapter 6 examines the continuing struggles about valid knowledge and appropriate standards of proof in the context of toxic tort litigation, an area of law that has proved as resistant to judicial rulemaking as to orderly legislation. Chapter 7 evaluates the role of the courts in shaping national policies for biotechnology, underscoring the judiciary's own adherence to prevailing views of technological progress. Chapters 8 and 9 review and contrast judicial performance in two areas of biomedical decisionmaking that have produced a spate of ethical dilemmas in recent years: the use of new reproductive technologies and life-sustaining technologies ("right-to-die" cases). Explanations are sought for the courts' apparently more successful ventures in the latter than the former cases, particularly along the dimensions of civic education and effectiveness. Finally, using the criteria outlined above, Chapter 10 reevaluates the dominant legal approaches to managing science and technology and presents some proposals for reforming a complex yet endlessly fascinating domain of legal practice.

2

Changing Knowledge, Changing Rules

Courts are no strangers to social conflicts arising from changes in technology. By the early seventeenth century, common law judges were already grappling with cases that pitted homeowners' interests against those of new commercial activities. Was a property owner entitled to damages if a "hog sty" was erected "so near the house of the plaintiff that the air thereof was corrupted" or a dye-house constructed so as to pollute his fish pond?[1] With the advent of industrialization, conflicts between old and new uses of land became more frequent, forcing courts to balance the economic benefits of technology against negative externalities such as noise, vibrations, smoke, and dust. Judges began to fashion remedies for landowners whose expectations with regard to their property were unreasonably threatened by technological development.

The pace of change quickened exponentially in the twentieth century, generating more complex problems for judicial decisionmakers, particularly in the United States. Courts, along with legislatures and administrative agencies, were called upon to cope with the social disruptions accompanying "our sudden, vast accelerations—in numbers, in the use of energy and new materials, in urbanization, in consumptive ideals, in consequent pollution."[2] Individuals, atomized and made more mobile by technology, found themselves caught up nonetheless in complex chains of production and consumption cutting across traditional boundaries of community and country.[3] The need for regulation grew stronger as the century advanced; yet, as U.S. industry struggled to hold its own in an increasingly competitive world economy, administrative regulation and tort liability imposed seemingly heavier costs in the United States than in countries with less liberal access to the courts. The risks of failing to regulate also loomed larger as the predictive power of science revealed problems of global dimensions, such as resource depletion and the reduction of biological diversity.

The legal system responded to these rapid changes in knowledge and technological capacity by adapting doctrines developed in the context of simple two-party litigation to the complicated market arrangements, social relations, and hidden risks of modern technological systems. Justice to the individual, the legal system's most basic yardstick, was reconceptualized in an environment in which, at one and the same time, scientific knowledge was becoming more complex and abundant and interpersonal relationships, together with notions of responsibility and blame, were undergoing rapid transformation. The story of the U.S. courts' emergence as policymakers for science and technology in the later twentieth century is primarily a story about the adaptation of the common law to the changed market conditions and regulatory needs of an advanced industrial society.

Doctrinal shifts occurred most noticeably, and most controversially, in those areas of the law in which technological change was bound up with profound changes in the public's expectations of liberty, privacy, and physical well-being. Disputes over products liability, medical malpractice, and environmental protection, for example, recast the courts as a proving ground for scientific claims supporting different theories of causation and responsibility. Collaterally, courts were called upon to decide whose knowledge to privilege when the personal and localized perceptions of lay individuals and communities came into conflict with the impersonal and universal assertions of science. The consequent changes in judicial attitudes toward proof, expertise, knowledge, and uncertainty form the subject of this chapter.

Products Liability

A recent two-page spread in a national newsmagazine identifies the liability system as a major threat to U.S. competitiveness and illustrates the point with a picture of an unnamed corporate research department with a "Closed" sign posted ominously in front.[4] Prominent legal analysts likewise maintain that courts have extended liability law far beyond its legitimate boundaries, thereby imposing an unseen tax on American society and depriving it of useful and essential products.[5] By what pathways did judicial decisionmaking stray into this thicket of hostile commentary?

Products liability matured relatively recently as a major branch of tort law. For much of the nineteenth century sellers of defective goods remained insulated from legal challenge by the general rule that they were liable only to those with whom they stood in a direct contractual relationship, termed "privity of contract."[6] Thus the consumer who acquired the product through resale from a dealer or other intermediate party had no recourse against the original seller, even if the seller's negligence had caused the injury. In 1916 Judge Benjamin Cardozo abandoned the ban on suits against sellers and manufacturers with his celebrated decision in *MacPherson v. Buick Motor Company.*[7] Cardozo's ruling quickly became the controlling doctrine regarding the obligations of manufacturers: by placing a product on the market, the manufacturer assumes responsibility not only toward the immediate buyer but also for any foreseeable harm caused to any subsequent purchaser. Rightly hailed as a progressive departure from the legal formalisms of the nineteenth century, this move aligned products liability law more closely with contemporary social conditions. Yet the decision remained true to the foundational principle of tort law that liability should be based on fault. Even after *MacPherson,* an injured consumer could recover only by proving that the injury was caused by negligence on the part of the manufacturer.

By the middle of the twentieth century, new intellectual currents were flowing across the field of torts. Under the influence of the Legal Realists, judges and academics became more sensitive to the potential of tort law as a policy instrument. The view that the primary function of tort litigation was to compensate injured persons gained ground over the countervailing doctrine that torts were mainly a legal device for punishing fault. The development of liability insurance during the early part of the century made it easier to separate the plaintiff's right to recovery from the defendant's blameworthiness. Insurance allowed the tort suit to be converted from a two-party dispute to "a 'three-party affair,' in which the third party was society at large."[8] Courts recognized that liability, particularly when covered by insurance, could effectively shift or spread losses away from individual victims without unduly penalizing commercial enterprises. By 1959 one influential commentator could approvingly claim that tort law was really "public law in disguise."[9]

Further doctrinal developments altered the complexion of products liability in the postwar world. As courts reduced their emphasis on fault, strict liability (liability without fault) replaced negligence as the primary basis for compensating persons injured by defective products. At the same time, courts began loosening some of the remaining threshold barriers to recovery for particular classes of plaintiffs, for example, workers injured in the course of employment or consumers injured by indeterminate third parties.

Justice Roger Traynor of the California Supreme Court, a distinguished inheritor of Cardozo's reformist mantle, played a notable role in the further transformation of products liability law. During Traynor's tenure as judge, the public policy implications of tort actions were more thoroughly explored and the notion of the manufacturer's responsibility was substantially redefined. Landmark cases grew out of unremarkable factual settings in which the cause of injury was never seriously in question. A waitress was injured when a Coca-Cola bottle exploded in her hand. A man working in a home workshop was badly hurt when a piece of wood suddenly flew out of a power tool and struck him on the forehead. Out of these everyday events, Traynor fashioned the doctrine that responsibility for injuries caused by defective products should "be fixed wherever it will most effectively reduce the hazards to life and health."[10] The producer, he noted, was in the best position not only to identify and correct the defect but also to distribute the risk of harm to the general public through appropriate insurance and pricing policies. Rather than promote costly and potentially fruitless inquiries into fault, Traynor argued, courts should reshape the law so that losses inflicted by hazardous products could be placed on the manufacturer—and so passed on to the consuming public.

The notion of strict liability stated in effect that factual certainty about the cause of an accident was sufficient to establish financial responsibility for loss. The California Supreme Court unanimously endorsed this principle in 1963 in *Greenman v. Yuba Power Products, Inc.*,[11] and it was soon adopted by other state courts. Within the next few years the concept of a "defective product" was expanded to include not merely products failing to meet the manufacturer's own specifications but also products, such as drugs and pesticides, that were defective either in their design or in their failure to carry adequate warnings from the manufacturer.

Meanwhile, the movement begun in *MacPherson* to extend the manufacturer's liability continued, bringing within the law's protection all those whose use of or exposure to the defective product was reasonably foreseeable.[12] This liberalization made good sense in light of the evolving judicial interest in compensating victims. Courts used the criterion of "reasonable foreseeability" to ensure that there would be enough connection between plaintiff and defendant to justify shifting the loss. As will be seen in Chapter 6, however, the doctrine of liability without fault, coupled with expansions in the classes of potential plaintiffs, exposed some industries (chemical manufacturers, in particular) to heightened risk of liability. Further, although courts positioned themselves increasingly as morally neutral risk spreaders, tort law continued to serve its traditional normative function by permitting high punitive damages in cases in which jury sympathy sided with victims. A dramatic example was the $105 million award against General Motors in early 1993 in a case involving a teenager's fiery death in a defectively designed GM pickup.

Another barrier that courts gradually whittled down in the

area of products liability was the requirement for plaintiffs to show that a particular defendant had caused their injury (the "determinate defendant" requirement). Courts noted that under modern circumstances of production and distribution it is often difficult or impossible for plaintiffs to identify the defendant with precision, even when there is little doubt as to which product or conduct caused the injury. The puzzle of how to fashion a remedy in such circumstances was posed early on in *Summers v. Tice,* a case involving a hunting accident.[13] Two hunters, both acting negligently, shot in the plaintiff's direction, and one injured him, but there was no way to tell which one was responsible. The court concluded that this fact alone should not bar the blameless plaintiff from seeking compensation. Under the rule of "alternative liability," the victim was permitted to recover from either of the two negligent marksmen. Once again, social uncertainty (which hunter was responsible) was resolved in the plaintiff's favor because there was unambiguous knowledge about the physical cause of the injury (the bullet used in hunting).

This precedent proved influential in later cases involving injury by toxic chemicals. Many chemical products, such as pesticides and pharmaceutical drugs, are made and marketed by more than one manufacturer and are, for all practical purposes, interchangeable. Injured parties may therefore find themselves knowing physically what harmed them without being able to specify who made the product in question. Thus, in another landmark California case, *Sindell v. Abbott Laboratories,*[14] the plaintiff could show that her injuries had been caused by the drug diethylstilbestrol (DES), administered to her mother during pregnancy, but she could not uniquely identify the responsible manufacturer. Instead of barring Sindell's claim, the court held that she could proceed against a group of manufacturers representing a substantial share of the DES market; if she succeeded, each defendant's liability would be determined according to its market share. Although other DES cases reached somewhat different results,[15] complex lawsuits since *Sindell* have continued to reduce the plaintiffs' evidentiary burden in cases involving many victims and multiple tortfeasors.

A relaxation of legal rules governing the responsibilities of employers loosed a flood of lawsuits in the 1970s by victims of workplace injuries and damage to health. Workers ordinarily are barred by law from suing employers for injuries arising from their employment; workers' compensation programs based on no-fault liability provide the basic remedy for work-related accidents and disease. This system was adopted in most states by the 1930s and remained essentially undisturbed for forty years.[16] But by the 1970s knowledge of latent occupational diseases strained the no-fault approach. Claims arising from such illnesses were often excluded by unrealistic time limitations, narrow listings of compensable diseases, and omission of diseases such as cancer or heart problems that also occur in ordinary life. Even when a claim was allowed, the worker had to prove that the condition was work-related. Trying to establish this link transformed a routine administrative settlement into a contentious, lengthy, and uncertain recovery process. The levels of compensation, moreover, fell pitifully short in relation to the gravity of many occupational diseases.

Frustrated and sick, workers began suing third-party manufacturers who had supplied disease-inducing products to their employers. A few important victories established products liability as a viable alternative route to compensation. Thus, asbestos insulation workers suffering from cancer and lung disease won the right to sue manufacturers in *Borel v. Fibreboard Paper Products Corp.,*[17] which applied the principle of strict liability to claims of health injury. A few successes by plaintiffs' lawyers led to a dramatic shift from workers' compensation to the tort system by asbestos claimants.[18] Although similar shifts occurred for other occupational diseases as well, the torrent of asbestos litigation remained unequaled in volume and impact. These claims alone contributed mightily to the perception that the American civil law system was suffering from an uncontrollable litigation explosion.

The cases that initially transformed liability law into an instrument of risk spreading took certainty about the physical causes of an event as justification for changing the rules regarding social responsibility. In accidents involving cars, lathes, or exploding bottles, and even in the DES and asbestos cases, courts were reasonably sure that deserving plaintiffs were being compensated with profits from harmful products. However, the doctrinal changes that expanded the plaintiffs' rights in these cases opened the courthouse door to cases in which the only certainty was the plaintiff's misery and ill health. In toxic tort cases, for instance, the cause of the victim's injury was much harder to establish, and disagreements over technical data were impossible to avoid. Thus, the liberalization of products liability law to compensate more victims encouraged litigation based on uncertain causal theories and drew seemingly insoluble scientific controversies into the courtroom. Put differently, a change in judicial policies for resolving social uncertainty—who should bear the cost of accidents—provided the impetus for addressing new kinds of scientific and technical uncertainty in court.

Medical Malpractice

In Brunswick, Georgia, a sleepy paper-mill town of only 17,000 people, a woman lawyer tried to arrange for the delivery of her third child. Local obstetricians refused to care for her, and the mother-to-be eventually had to drive eighty miles to Savannah to have the baby delivered.[19] The boycott was in retaliation for a malpractice suit that the woman had filed against an obstetrician in her own town. This small personal drama enacted in an out-of-the-way part of the country in 1986 was symptomatic of a much broader confrontation between the medical and legal professions that took shape a decade earlier. In the mid-1970s successful malpractice claims led to policy cancellations and sharp increases in insurance premiums as nervous insurers pulled out of the medical market. From 1974 to 1976 alone, premiums for some specialties rose by several hundred percent; insurance rates for hospitals doubled and in some states even tripled.[20] Medical malpractice became a major public issue for state legislatures and even the U.S. Congress.

Did the surge in medical malpractice claims during the 1970s reflect fundamental changes in judicial attitudes toward the scientific and medical professions? Did malpractice cases inappropriately thrust courts into second-guessing medical experts? And what policy approaches evolved for minimizing conflicts between the legal and medical professions? The history of infectious disease control in the nineteenth and twentieth centuries provides a starting point for addressing these questions.

Deborah Merritt's study of judicial behavior from 1875 to 1910 indicates that courts a century ago generally deferred to medical experts and health departments when faced with challenges to public health measures.[21] The Supreme Court of Pennsylvania's reluctance to intervene in an "abstruse question of medical science" in a smallpox vaccination case typified the hands-off position adopted by judges.[22] A hundred years later, a more skeptical attitude could be discerned, particularly when public health had to be balanced against individual liberty. In a 1978 New York case, for example, a federal court was not satisfied with the state board of education's claims that several dozen mentally retarded carriers of hepatitis B should be segregated in separate classrooms.[23] The court required the board's medical experts to demonstrate actual harm rather than the mere probability of transmission. This ruling held the experts to a higher standard of proof than the experts themselves had deemed necessary.

Malpractice claims still more directly substituted legal standards of appropriate conduct for those of medicine. In malpractice suits, patients may question a doctor's actual performance, the choice of a treatment strategy, or the adequacy of information provided about the selected treatment. A key issue in deciding whether a defendant physician has acted properly is the standard against which the challenged conduct should be measured. In one controversial line of cases, courts changed the long-standing rule that doctors should be held only to the standards of medical care prevalent in their local communities.[24] Given the increasing specialization of medicine, courts concluded that the performance of a practitioner anywhere in the country might appropriately be compared with that of other professionals in the same specialty, regardless of their location. Nationalizing the standard of acceptable medical practice also provided a safeguard against the professional solidarity that prevented doctors practicing within a given locality from testifying against one another.

But the abandonment of the locality rule had unexpected consequences for the culture of malpractice litigation. It encouraged a cadre of "career" expert witnesses to appear on behalf of plaintiffs in any jurisdiction.[25] With a national pool to draw on, plaintiffs' attorneys could more easily locate a witness to argue that the defendant had not read a crucial article or should have taken more precautions.[26] Juries knowing little of the intricacies of medical judgment were often at sea in deciding whether the alleged deficiency was trivial or a sufficiently serious departure from accepted practice to constitute negligence. Lobbied by physicians, many states undid the judicially imposed national standards with laws restoring a community, state, or local standard in malpractice cases.[27]

Some courts also rejected the test of local medical convention in evaluating the adequacy of informed consent. Starting in the 1960s, courts began to impose on doctors an affirmative duty to inform patients of the risks associated with their treatment procedures. Professional norms within the medical community determined which risks should be disclosed and how much information should be provided. But this approach ran into difficulties when experts could not agree on the criteria for informed consent and juries had to make their own decisions about which expert to believe. In the 1970s some courts held that the lay patient rather than the doctor should be the reference point for deciding whether the information given was adequate.[28] The jury could then be asked whether the doctor had provided all the information that a patient would reasonably wish to have. Jethro Lieberman argued in his study of American litigiousness that this development substituted "a legal for a medical model of man": "Judges trained in the common law are not apt to think of an individual as a biological organism whose care is best entrusted to a medical professional. Rather, the judge will consider the individual as an autonomous being with rights, among them the right to make up his own mind."[29] Perhaps more to the point, the new standard accorded the jury's lay opinion more weight than that of medical experts,

possibly because judges had faith in the jury's ability to balance expert views about the adequacy of information against
the patient's needs as an autonomous, rights-bearing individual. State legislatures, however, again came to the profession's
rescue. In New York, for example, a state law overruled a judicial holding that the patient's point of view should govern in
disputes concerning informed consent.[30]

In one celebrated case, a court in Washington state extended
the principle of liability without fault, normally applied only to
manufacturers of defective products, into the area of medical
malpractice.[31] A woman of thirty-three sued the ophthalmolo-
gist who had been treating her for ten years for failing to detect
and control the glaucoma that had severely and permanently
damaged her eyesight. Evidence at trial showed that early detection and treatment could have saved her vision, but that the
doctor had followed applicable medical norms. Glaucoma tests
were not commonly administered to patients under forty, since
the disease so rarely occurs in younger people. Nevertheless,
the Washington Supreme Court unanimously held that the
doctor had been negligent in failing to give the test. As in the
products liability cases, the decision seemed to be driven by
sympathy for an innocent plaintiff and the clearly preventable
nature of her injury.

Although critics predicted disaster for the medical profession, this draconian decision had little influence on malpractice law in Washington or elsewhere. The state legislature restored the traditional negligence standard a year after the
supreme court decision, requiring plaintiffs to prove as before
only that the defendant had failed to exercise the degree of care
possessed by "other persons in the same profession."[32] Some
years later, the Washington Supreme Court showed that it was
prepared to sidestep even this legislated standard to compensate another glaucoma victim whose doctors had failed
through routine testing to detect her high risk of disease.[33]

The history of medical malpractice after 1970 thus presents
a paradoxical picture. On the one hand, courts initiated a
wide-ranging retreat from nineteenth-century attitudes of deference to medical expertise; in the process, they created a hospitable environment for submitting medical claims to critical

legal scrutiny. By pushing for national standards of care, judges asserted their right to determine the applicable codes of conduct for the medical profession and to equalize these across the country. As in contemporaneous public health controversies, courts showed greater concern for individual rights and lay standards of reasonable care than for the medical community's professional autonomy and judgment. The rule that informed consent should be evaluated from the patient's standpoint was a notable victory for the nonexpert perspective. On the other hand, following the malpractice crisis of the mid-1970s, state legislatures reversed many of these judicial incursions into malpractice law, underscoring both the power of professional medicine and the lack of organized political support for the positions taken by the courts.

Further complicating the picture, state tort law reforms had little detectable effect on the volume of malpractice litigation. A study conducted by the General Accounting Office (GAO) for the U.S. Congress in 1986 indicated no agreement that any of fourteen major reforms of the tort system had noticeably influenced litigation patterns, although some interest groups felt that caps on awards and the use of pretrial screening panels to eliminate frivolous claims may have been effective to a limited degree.[34] These findings support the thesis that judicial involvement in refashioning policies for science and technology can have repercussions long after courts cease actively to make modifications in the law.

Although legislatures reined in many liberalizations introduced by the courts, the sympathy shown to suits against physicians permanently affected the social context of litigation. Malpractice claims were demonstrated to be both legitimate and winnable. The culture of expert witnesses changed with the elimination of the locality rule, making it more acceptable for physicians to testify against other physicians. More important, the fact that some patients began to win malpractice cases created incentives for others to seek redress; as in the area of asbestos litigation, only a few key successes were needed to alert plaintiffs, and their attorneys, to a significant window of opportunity. Programs for using expert panels as an alternative to litigation proved to be such a bottleneck that they were struck down as unconstitutional in half a dozen states, including Pennsylvania. Economic and social factors thus sustained the momentum created by the courts. And the controversies unleashed by judicially created changes in liability rules propelled courts as never before into reviewing the *substance* of decisions by the medical community.

Environmental Litigation

About a hundred years ago, on the banks of the River Irwell near Manchester, England, a paper manufacturer was carrying on a business for which he needed very pure water. Six miles upstream, an alkali manufacturer began depositing heaps of refuse on a piece of land close to the river, threatening to pollute the water required by the downstream user. The paper manufacturer, whose livelihood was thus endangered, asked for an injunction on the grounds that "in the course of a few years a liquid of a very noxious character would flow from the heap, and would continue flowing for forty years or more, and that if this liquid should find its way into the river to any appreciable extent the water would be rendered unfit for the Plaintiff's manufacture, and his trade would be ruined." The court, in *Fletcher v. Bealey,* refused the request because the danger to the plaintiff was not imminent: the facts did not meet the common law test that harm must be "practically certain to occur" before a court would order the harmful conduct to cease. Explaining his decision not to act, the judge commented, "I think that in ten years time it is highly probable that science (which is at work on the subject) may have discovered some means for rendering this green liquid innocuous."[35]

This faith in the power of science echoes quaintly from the Progressive Era to a contemporary America where predictions of irreversible and catastrophic damage to the environment dominate political discourse, and where public interest groups routinely enroll the courts in their efforts to protect the environment. Yet it was not until the late 1960s that U.S. courts began expressing any misgivings about remote and speculative risks to health and the environment. Judicial activism on behalf of the environment coincided with the rise of environmental consciousness in the public, fed by the same forces of scientific and social change.

Steadily accumulating knowledge about the long-term effects of pollution, produced partly by research and partly by accident, was one factor that reshaped judicial thinking on environmental issues. The list of triggering events seemed unending: the disastrous impact of polychlorinated biphenyls (PCBs) on aquatic life, of DDT on birds, and of acid rain on forests; the health hazards of chronic, low-level exposure to lead; the contamination of groundwater by pesticides; and the depletion of stratospheric ozone by chlorofluorocarbons. The second factor was the substitution of a statutory for a common law approach to regulating polluting industries. Congress enacted a spate of legislation aimed at identifying environmental risks in advance and controlling them before they caused harm. The National Environmental Policy Act (NEPA) of 1969, in particular, occasioned a revolutionary change in the role of the courts. Almost overnight the courts became a battleground for competing environmental ideologies as activists sought judicial aid in enforcing NEPA's mandate to incorporate environmental values into governmental decisions.

Judicial review emerged in the 1970s as a formidable instrument for forcing action by torpid governmental agencies and blocking or delaying unwelcome industrial development. The courts opened their doors to environmentalists, arguably even wider than Congress had intended. For instance, the judiciary embraced an active role in interpreting and enforcing NEPA, although neither the statute nor its legislative history explicitly provided for judicial review.[36] In subsequent legislation Congress made it plain that courts should indeed serve as watchdogs over environmental regulation. From 1970 to 1980 most new laws for protecting the environment gave courts the power to review policies developed by administrative agencies. They also authorized private citizens to enforce environmental regulations in court when the responsible agency had failed to take appropriate action. By 1972 the courts were so firmly established as friends of environmental protection that the legal scholar Christopher Stone broached the provocative (if only academic) idea of giving trees as well as people standing to defend their rights in court.[37]

Environmental legislation in the 1970s fundamentally restructured the judicial approach to considering scientific evidence of risk. Statutes such as the Clean Air Act and the Clean Water Act instructed the government to regulate activities that endangered public health. With the approval of federal judges, agency officials interpreted the term "endanger" as carrying a preventive meaning, requiring attention to unproved future harms as well as demonstrable present ones. Any risks that could reasonably be surmised on the basis of available scientific evidence became legally significant, and regulators were empowered to intervene under conditions of lesser certainty than prescribed by the common law standard of "imminence."

Two landmark environmental decisions of the 1970s spelled out the implications of precautionary legislation for scientific proof: *Reserve Mining v. EPA*[38] and *Ethyl Corporation v. EPA*.[39] In *Reserve Mining* the U.S. Court of Appeals for the Eighth Circuit had to consider whether discharges from the mining company's processing of low-grade iron ore (taconite) posed a sufficient threat to public health to justify remedial action. The wastes, or "tailings," from the taconite operations resembled asbestos fibers in structure, causing concern that their discharge might expose the public to a risk of cancer comparable to that from exposure to asbestos.

The evidence, however, was far from conclusive. Were taconite fibers sufficiently similar to asbestos to justify concern about their carcinogenicity? Did the length of the fibers matter in estimating the degree of risk? Epidemiological studies had shown that inhaling airborne asbestos fibers caused cancer. Would ingesting taconite fibers in drinking water drawn from Lake Superior present similar risks? What were the expected levels of human exposure to the taconite, and was the risk at these exposures significant enough to merit legal attention? Given the equivocal and incomplete nature of the evidence on all these points, the court had to find a new principle for deciding under scientific uncertainty. The Eighth Circuit found a precedent in a dissenting opinion written by Judge J. Skelly Wright of the District of Columbia Court of Appeals in *Ethyl Corporation v. EPA*, a contemporaneous case under the federal Clean Air Act that also hinged on the statutory concept of "endangering" public health. Key to Wright's analysis was the notion that "endanger" meant "something less than actual harm."[40] To meet this standard, Wright argued, one had merely to show that harm was threatened, not that it would actually occur. The Eighth Circuit concluded that the facts in *Reserve Mining* satisfied this interpretation of endangerment. An acceptable, though unproved, medical theory provided the grounds for "a reasonable medical concern over the public health."[41] Accordingly, the court ordered Reserve to abate the pollution within a reasonable period by switching to land disposal of its wastes.

The influence of *Ethyl* reached well beyond *Reserve Mining*. In a second *en banc* hearing, the D.C. Circuit reversed its earlier position. Judge Wright, writing this time for the majority, again emphasized that conclusive evidence was not needed in order to take regulatory action under a precautionary environmental mandate. Danger, Wright cautioned, does not correspond "to a fixed probability of harm, but rather is composed of reciprocal elements of risk and harm, or probability and severity. That is to say, the public health may properly be found endangered both by a lesser risk of a greater harm and by a greater risk of a lesser harm."[42] In the years since *Ethyl*, risk assessment for regulatory purposes has developed into an incomparably more complex and controversial technique than Judge Wright could possibly have foreseen when he asked decisionmakers to weigh risk against harm. However, his insight that risks to public health and the environment may be regulated on the basis of a reduced quantum of proof remains a foundational principle of American environmental policy.

Continuity and Change

Historical developments in three areas of the law illustrate the pathways by which U.S. courts established a heightened receptivity toward controversies involving new kinds of scientific and technical arguments. In each area, the changes were driven by a different mix of judicial innovation, legislative prescription, and new understanding of biological or social relationships. But changes in all three areas of the law corresponded to more general trends in the relations among science, technology, and society. The predisposition of the courts to hear disputes based on uncertain evidence and to allow lay plaintiffs to challenge scientific professionals reflected the continuing erosion of public trust in the governability of science and technology. The skepticism voiced by lawyers and judges in the courtroom echoed a more widespread disenchantment with expert assurances about risk. As the primary custodians of individual rights, courts were sensitized to threats posed by science and technology to individual safety and autonomy. Concerns for promoting public participation in technological decisions and protecting individuals against involuntarily assumed risks emerged as important themes in legal controversies over science and technology.

The look of controversy in each area of law changed again in keeping with the political changes of the 1980s and 1990s. Among the notable shifts were a rise in complex litigation occasioned by the consolidation of large products liability claims and a pronounced tilt toward protecting private property owners in environmental disputes with government agencies. Even in these altered circumstances, however, courts continue to grapple with many of the same normative questions: whose knowledge counts in defining new legal rights and responsibilities; who should be believed when competing technical arguments are in play; and how much certainty should be required in order to rationalize changes in the rules of compensation or environmental stewardship? Two decisions that exemplify these lasting preoccupations can be read as a coda to this chapter and as a bridge to the next.

In *Criscuola v. Power Authority of the State of New York*,[43] the New York Court of Appeals had to decide what proof was needed in order to compensate a property owner for an alleged drop in property value adjacent to a right-of-way for a high-voltage power line. The putative reason for the loss in market value was the public's fear of cancer and other health damage from electromagnetic fields around power lines. The court ruled that the plaintiff had to show that fear existed and had diminished the property's market value, but not that the fear was reasonable. Evidently, the court felt that expert testimony from economists testifying about land prices would be easier for the legal system to process than testimony from the less familiar lineup of "a whole new battery of electromagnetic power engineers, scientists or medical experts."[44]

In *Dolan v. Tigard,*[45] the U.S. Supreme Court was confronted with a question that was not on its face highly scientific. Were the environmental restrictions that the city of Tigard, Oregon, had imposed on Mrs. Florence Dolan's plans to expand her store so unreasonable as to constitute an uncompensated "taking" of property under the Fifth Amendment? Ruling for Dolan by a bare five-to-four majority, the Court announced a standard that could have serious consequences for proof and evidence in other similar cases. The city, according to Chief Justice Rehnquist's majority opinion, had to demonstrate that the conditions attached to the permit were in "rough proportionality" with the development's likely impact. Although Rehnquist stated that "no precise mathematical calculation is required" to make the necessary showing, John Echeverria, general counsel of the Audubon Society, observed that the ruling was likely to make land-use decisions more expensive and time-consuming, since both local governments and developers would have to do more studies and submit more information.[46] That this requirement in turn would increase the probability of disputes over risk and cost assessment was not stated but can be confidently surmised from the entire recent history of environmental regulation.

3

The Law's Construction
of Expertise

The legal system has long looked to science as an indispensable ally in a shared project of truth-finding. A talmudic story from the sixth century B.C. illustrates the law's early dependence on scientific expertise: "A husband, desiring to divorce his wife, contrived to get her and other guests drunk at a party, carried her and a male guest to a couch and threw egg albumen between them. He then called the neighbors to bear witness to adultery."[1] The resourceful wife became sober and called in her physician, who identified the substance as egg white, not seminal fluid. We are not informed what tests the doctor used or whether the husband sought to escape his predicament by resorting to any form of counterexpertise. Perhaps the conflict ended because no one thought to question the medical expert's positive identification.

By the middle of the fourteenth century, the practice of obtaining specialized information from skilled persons was relatively commonplace in the courts of England, although the functional separation between jurors and witnesses was not firmly established until the sixteenth century.[2] Experts in these early cases were frequently called by the court, but the common law system of letting the parties summon their own witnesses was well in place by the eighteenth century.[3] Rules governing the admissibility of expert evidence were also taking shape during this period. In a leading English case of 1782,

Lord Mansfield permitted an engineer to give his opinion about the cause of a harbor's filling up, noting that "the opinion of scientific men upon proven facts may be given by men of science within their own science."[4] The use of expert witnesses was widespread in American legal proceedings by the turn of the century. In 1909, by one account, 60 percent of the cases tried in a Massachusetts superior court used some form of specialized testimony.[5]

With the surge in technology-related lawsuits in the past few decades, there has been a dramatic increase in the number and variety of experts participating in U.S. legal disputes. Witnesses representing psychiatry, sociology, statistics, geology, epidemiology, and toxicology, as well as more novel and arcane disciplines such as linguistics and the philosophy and sociology of science, have entered the courtroom alongside the more familiar figures of engineers, economists, physicians, and forensic scientists. Yet legal institutions and procedures for dealing with technical evidence have remained remarkably static. Most U.S. judges are still generalists, without any special schooling in the sciences, and practices such as random assignment of cases prevent judicial specialization in areas requiring technical knowledge. Juries, who are responsible for a high percentage of legal fact-finding, represent an even lower level of scientific sophistication. The average American juror has attained no better than a high-school education. As a result, the costly ritual of examining witnesses in the courtroom is directed mostly toward translating, repackaging, or—when the occasion demands—debunking highly technical information for persons whose grasp of basic scientific principles is negligible or nonexistent.

More important, critical thinking about the role of science in the legal process has advanced little beyond Lord Mansfield's simple faith in the power of experts to facilitate fact-finding. Most legal practitioners are aware that the law's own testing and establishing of "facts" is always contingent on the goals of resolving disputes and dispensing justice. Yet systematic studies of the contingency of scientific knowledge and its implications for law remain to be undertaken. As we shall see, legal fact-finding is constrained not only by the moral and ethical standards articulated by judges but also by sociological factors specific to both law and science, including the ways in which expert witnesses are selected and acculturated for courtroom appearances, as well as the general processes by which claims gain authority within the scientific community. As a result, the legal fact-finder receives accounts of science that are contingent on the circumstances of the particular case and imbued with the parties' particular normative agendas. Within the courtroom, each side's elaborately constructed expert testimony is subjected to an adversarial process whose primary function is to expose the contingencies in the offering party's evidence, thus undercutting its claim to credibility.

Preparations for the murder trial of O. J. Simpson through much of 1994 offered a remarkable public window on the construction of scientific evidence as part of a more encompassing strategy for establishing the defendant's guilt or innocence. In this no-holds-barred, no-expenses-spared confrontation, disputes over the reliability of "DNA fingerprinting" (more properly known as DNA typing)—a sophisticated identification technique—erupted long before the trial. As each side attempted to discredit the other's credibility, science was enlisted as one resource among many. Complicating the picture, the scientific community tried to carve out an independent role through such means as committee reports by the National Academy of Sciences and articles in *Science* and *Nature.* But these episodes only brought home the point that science in a legal setting is always bound up with specific constructions of causation, blame, and responsibility.

In this chapter we shall see in detail how expert claims are presented and deconstructed in the courtroom, and how courts ultimately achieve closure in spite of the adversarial structure of litigation. The law over time has evolved numerous formal rules and procedures to ensure the reliability of scientific expertise, but of at least equal interest are the informal practices and techniques by which courts certify the facticity of some claims and deny the validity of others. Surveying these formal and informal systems for screening knowledge, we will ask which approaches to institutional or procedural reform are most likely to enhance the effectiveness and educational capacity of the courts.

The Cultures of Expert Witnessing

Martin Shapiro, an authority on comparative legal systems, attributes to Anglo-American law a special penchant for fact-finding, which he relates to its preference for binary, right/wrong decisions. "Even where facts cannot be established with certainty," Shapiro notes, "Anglo-American courts typically make definitive findings of fact and treat them as certain even though they are established only by the preponderance of evidence."[6] In France, by contrast, judges are more willing to recognize the limits of fact-finding, using presumptions when necessary to bridge the gaps in the evidence.

Ironically, however, it is the common law tradition, with its exceptional devotion to "finding the facts," that has dispensed most completely with the notion of the neutral expert. Expert witnesses in U.S. courts are sought out, trained, and compensated by the disputing parties. Unlike civil law tribunals, the judge in a classic common law proceeding plays little or no part in managing the production of evidence, structuring its presentation, or examining witnesses, except to exclude material that is deemed inadmissible by law. What the common law adjudicator sees in practice are two carefully constructed representations of reality, each resting on a foundation of expert knowledge but each profoundly conditioned by the culture of expert witnessing as it intersects with the interests, ingenuity, and resources of the proffering party.

The Commodification of the Expert

As technical evidence becomes an increasingly valuable commodity in legal controversies, lawyers and their clients have become acutely conscious that its value rises if it can be presented as the truth or the best approximation to it.[7] Leading expert witnesses command skyrocketing prices, limiting access for poor and uneducated plaintiffs. In 1983, for instance, a Manhattan psychiatrist and neurologist was said to be earning as much as $200,000 from his work as an expert witness and legal consultant.[8] Four years later, fees for expert witnesses reportedly ranged "from $50 an hour for a law-enforcement expert to more than $10,000 a day for a plastic surgeon."[9] While some experts may offer their services for modest fees, and some may even testify *pro bono,* these figures underscore the sober commercial calculations that govern the use of expertise in the legal system. Inequality of resources in any lawsuit can easily tilt the balance in favor of the wealthier party.

Expertise may be "owned" outright by partisan actors. For example, law firms that deal with malpractice claims frequently employ an in-house staff of medically trained personnel to review and prepare new cases. More often, expertise is bought for a specific purpose. Both the defense and the prosecution in the O. J. Simpson case scoured the country to locate the most authoritative scientific allies they could muster on the issue of DNA typing. The prosecution contracted to have its DNA testing done by Cellmark Diagnostics, a highly regarded private firm, while the defense scored a coup by enlisting not only a stellar cast of lawyers but also the head of Connecticut's crime laboratory, a forensic scientist of unimpeached integrity.

For lawyers in routine cases, the search for the right expert witness is increasingly dominated by a variety of middlemen, either witness brokers who specialize in finding experts for particular types of lawsuits,[10] or clearinghouses, such as the Expert Witness Network and the Technical Advisory Service for Attorneys. Despite their obvious advantages, such intermediaries create entry points for unethical or professionally marginal experts, as in the case of a physician who became a legal consultant after serving a prison term and losing his license to practice medicine.[11]

In the commodity market of expertise, persuasiveness more than raw scientific credentials determines a witness's worth. Experts may seek to establish themselves (often with the help of entrepreneurial lawyers) as specialists in particular types of cases, sometimes appearing categorically as "plaintiff's witnesses" or "defense witnesses." The legal system's preference for proven winners encourages such repeat witnessing, although it substantially narrows the range of expertise that finds its way into court.[12] Thus Dr. Bertram Carnow, a professor of environmental and occupational medicine at the University of Illinois, testified in 1982 in a trial that produced a $58 million jury verdict for railroad workers who claimed to have been injured by dioxin. Two years later Carnow was sought out by a New York attorney who needed a dioxin expert to testify for the Vietnam veterans in the Agent Orange case. Subsequently Carnow emerged as a leading plaintiff's witness in cases involving claims of immunological damage from exposure to chemicals.[13] In an illuminating study of the products liability trials involving the drug Bendectin, Joseph Sanders, a specialist in the sociology of law, notes that expert witnesses for the plaintiff testified on average in more than ten cases while defense experts testified in an average of seven.[14]

The Pressures of Advocacy

Law in theory shares with science the "effort to be wholly rational, to organize and institutionalize the search for truthful data, and, above all, to seek truthful data as the basis for judgments."[15] Legal practitioners who subscribe to this positivist vision of the law regard the adversary system as an unfortunate corruption of an ideal state of truth-seeking. Having experienced litigation as both judge and advocate, Marvin Frankel echoed this disenchantment: "We know that many of the rules and devices of adversary litigation as we conduct it are not geared for, but are often aptly suited to defeat, the development of the truth . . . Employed by interested parties, the process often achieves truth only as a convenience, a by-product, or an accidental approximation."[16] Frankel concluded that "truth and victory are mutually incompatible for some considerable percentage of the attorneys trying cases at any given time."[17] Peter Brett, another critic of the adversary process, expressed the same sentiment. "At a common law trial," he noted, "there is no real reason for supposing that the truth will emerge; for the adversary system is designed so as to permit the parties to conceal it from the court."[18]

Lawyers no less than scientists, it seems, have a stake in representing the adversary process as a fall from grace, a system that subverts and distorts truth for the sake of winning. Critics are often acute in identifying reasons for this distortion, although their accounts invariably locate the problems within the law. Thus, adversary proceedings are thought to be inherently incompatible with the idea of telling the "whole truth." An expert who finds that she has conveyed a misleading impression to the court may not be able to correct it if counsel fails to ask the right follow-up question.[19] Telling "nothing but the truth" is equally farfetched in practice. Any lawsuit opens up endless possibilities for technical experts to stray beyond their acknowledged domains of competence. Paul Meier, a statistician at the University of Chicago, calls attention to the problem of "aggrandizement": the temptation of experts to give definitive rather than qualified answers, to deemphasize the existence of other schools of thought, and to exaggerate the significance of their own inferences.[20] Litigation, more generally, offers no hard-and-fast rules for separating facts from scientific opinions, opinions from mere speculation, or speculation from legal conclusions. The borderline between value decisions that should be the prerogative of judges and factual matters that are properly reserved for experts is itself a construct, negotiated anew from expert to expert and from case to case.

Practical guidelines on expert witnessing that have proliferated in professional journals show that would-be experts are well aware of the opportunities for strategically framing testimony in a legal forum. A guide to engineers instructs its readers: "Your presentation of testimony should generally follow one of two methods. One is to set forth all evidence bearing on the case. The other is to withhold some nonessential evidence which is damaging to the opposition, anticipating that opposing counsel will 'rise to the bait' and request it during cross-examination . . . This delayed introduction effectively strengthens your testimony and diminishes opposing counsel's enthusiasm for further questioning."[21] Even if the adversary is a friend or colleague, one is "obligated to make little of him and to destroy his opinion (rightfully)" by one's own superior knowledge.[22]

Within the legal culture, expert participants are evidently prepared to establish their credibility through tactics that have little to do with establishing scientific competence among a body of professional peers.

The Reluctant Expert

Scientists are often unwilling to play the expert witness game in accordance with the priorities and commitments of professional lawyers. The open admission of strategic goals in litigation cuts against the grain for people who are trained to think of themselves as impartial observers. The rules of legal argumentation compound their unhappiness. Not every scientist is willing to decide between two rival theories, knowing "(a) that neither of the rival theorists was bound to put forward all the data in his possession—indeed, that each would regard it as proper to suppress any 'inconvenient' or inconsistent observations of whose existence he knew and (b) that . . . the adjudicator, would be precluded from suggesting or requiring the elicitation of any additional data that might prove critical."[23] For many potential expert witnesses, these ground rules present an insurmountable ethical obstacle to participation in the legal process.

Another powerful deterrent is the belief that public disagreement with one's peers brings discredit on one's discipline. The issue arose in a challenge to an Arkansas law requiring public schools to give balanced treatment to "creation-science" and "evolution-science."[24] Michael Ruse, a historian and philosopher of science, had testified on behalf of the plaintiff, the American Civil Liberties Union. Philip Quinn, another philosopher, criticized Ruse's testimony on the ground that it was not "representative of a settled consensus of opinion in the relevant community of scholars."[25] Ruse argued back that an expert's views need only be "reasoned and based on one's knowledge of one's field."[26]

Practitioners in numerous scientific and technical disciplines—notably, psychiatry,[27] medicine, and, more recently, statistics[28]—have had to contend with negative publicity to their field from frequent entanglements with the law. The spectacle of psychiatrists quarreling over predictions of a criminal defendant's dangerousness, for example, weakens public trust in the reliability of the profession's expert judgments. It is hardly surprising, then, that the majority of scientists prefer to air their differences in the safe havens of professional meetings and journals, where a discipline's authority and internal cohesion can be more easily maintained. For lay citizens, however, the unpredictable rough-and-tumble of the adversary process offers the only form of engagement that renders transparent otherwise invisible cultural or normative commitments embedded in experts' claims about the "real world."

Science Made for Courts

Unsung in most academic writing on science and law is the growing influence of legal proceedings on the production of new scientific knowledge and techniques. Often research is undertaken only when a lawsuit points to the existence of a previously unsuspected causal connection, such as electromagnetic fields and cancer or silicone gel breast implants and immune system disorders. Research in these instances develops at best in parallel with legal proceedings but more often may lag behind them. As a result, decisions may well be taken before the scientific community has generated data reliably confirming or failing to confirm a particular causal claim. In June 1994, for example, the *New England Journal of Medicine* published a retrospective epidemiological study of 749 women who had received a breast implant but who displayed no statistically significant increase in connective tissue diseases. The news came months after Dow Corning Corporation, the nation's largest manufacturer of breast implants, had announced a multimillion-dollar settlement with claimants. Marcia Angell, the *Journal*'s executive editor, seized on the occasion for a general attack on the legal system's handling of scientific evidence: "scientific conclusions cannot be based on argument and opinion. There must be data. Yet, in the courtroom, acceptance of expert testimony usually turns on the 'credibility' of the witness, not the validity of the evidence on which the witness's opinion is based."[29]

Angell's naive historiography, which assumes that judges and juries could in principle have had access to "data" in this case, ignores both the agency of the law in producing relevant scientific knowledge and the lapse in time between the law's need for knowledge and science's ability to provide it. George Annas, another normally sophisticated observer of law and science, similarly fell victim to oversimplification when he deplored the fact that, for DNA testing, "standards of admissibility in the courtroom and the weight of evidence were being debated in the scientific literature whereas the standards of scientific validity were being debated at the same time in the courtroom."[30] In endorsing the view of some scientists that the problem arose from an overly hasty introduction of science into the courtroom, Annas bought into a historically and sociologically naive accounting of the pathways by which this particular forensic technique gained its legal standing. Peter Huber's more polemical work, *Galileo's Revenge,* is also marred by careless histories of the relationship between the growth of knowledge and the law's need for and use of it.[31]

When scientific expertise is produced in response to litigation, science's normal processes of validation can be bypassed or distorted. DNA typing evidence compiled by private testing laboratories was used in dozens of U.S. criminal trials, although scientists later characterized some of these practices as unacceptable and even indefensible.[32] Techniques such as visual imaging of brain function through positron emission tomography (PET) scans or computer animations "reconstructing" crimes and accidents are used with little attempt to validate them or to explain why they should be interpreted in particular ways.[33] Peer review standards for courtroom science are either nonexistent or may evolve in ad hoc fashion reflecting scientists' own perception of what the law requires, as when the editor of the *Journal of the American Medical Association* submitted an article to extra review because it had significant implications for litigation.[34]

Scientists have recognized as well that research and publi-

cation, especially peer-reviewed publication, are important resources in litigation. The *Nature* article on DNA typing by Eric
Lander and Bruce Budowle was interpreted by some as an
attempt to create the appearance of consensus and thus to
provide easier entry for DNA testimony in the O.J. Simpson
murder trial.[35] In other litigation contexts, scientists with particular interpretations of the evidence have sought out friendly
journals to publish, and thereby enhance the credibility of,
their claims, with an eye to future litigation.

The Deconstruction of Adversary Science

Defenders of the adversary process claim that legal decisions
are fairest when two parties argue "as unfairly as possible,
on opposite sides, for then it is certain that no important
consideration will altogether escape notice."[36] Indeed, cross-
examination has been lauded as "the greatest legal engine ever
invented for the discovery of truth."[37] These observations
reflect the legal system's persistent commitment to the view
that a trial is an occasion for locating the truth rather than for
choosing between alternative constructions of possible realities.

Research in the sociology of scientific knowledge suggests a
very different reading of what happens when cross-examination
and other adversary procedures are brought to bear on
scientific representations. Contradicting the logical positivists'
view that science simply mirrors nature, this body of work
stresses the social factors that go into the production of
scientific knowledge.[38] The authority of scientific claims derives, according to the sociological account, not directly, from
the representation of physical reality, but indirectly, from the
certification of claims through a multitude of informal, often
invisible, negotiations among members of relevant disciplines.
A complex network of people, methodologies, visual recordings
or inscriptions, and instruments (which themselves incorporate social conventions)[39] must be brought into harmony in
order to establish scientific claims as true. The facts of science
may over time become so widely accepted that it is no longer

possible to see how they were originally put together. But even when a claim's factual status is still "in the making," its provisional nature may be screened from public view by a technique that sociologists of science call "boundary work": that is, a communally approved drawing of lines between "good" and "bad" work (and, not trivially, between good and bad *workers*) within a single discipline, between different disciplines, and between "science" and other forms of authoritative knowledge. Boundary work, as we shall see, occurs just as commonly within the legal system when fact-finders make distinctions between valid and invalid presentations of evidence.

One may reasonably expect the constructed character of science to become apparent when claims are subjected to deliberate hostile scrutiny, as under cross-examination. Indeed, studies of U.S. regulatory proceedings have documented that scientific claims are inevitably deconstructed, disclosing areas of uncertainty and interpretive conflict, when evidence is developed in accordance with adversarial procedures.[40] There is, however, a bias toward maintaining the institutional authority of science even in these orgies of deconstruction; legal fact-finders are as committed to showing that truth exists as the scientists whom they interrogate on the stand.

Tests of Credibility

The adversary process shows to greatest advantage when it catches the expert witness in an outright lie, but such exhilarating moments are more familiar to aficionados of televised courtroom drama than to real-life practicing attorneys. Thus, in *Berkey Photo, Inc. v. Eastman Kodak Co.*,[41] a well-known antitrust case, the lawyers for Berkey unexpectedly came across a letter from Kodak's sole expert witness, the economist Merton J. Peck of Yale University, stating that Kodak's past anticompetitive practices had contributed to its current market strength. Berkey's attorneys very effectively used the discovery to establish that Peck had knowingly concealed evidence, therewith demolishing both his credibility and Kodak's case. Gordon Stewart, a plaintiff's witness in a case involving whooping cough vaccine, was similarly discredited when cross-examination revealed that he had misrepresented the nature of the controls in a major epidemiological study.[42]

All the factors that go into establishing a witness's credibility—not only knowledge but also social and cultural factors such as demeanor, personality, interests, and rhetorical skills—are simultaneously open to attack when scientific testimony is subjected to the adversary process. Critics accordingly worry that the nonepistemological determinants of credibility may carry the greatest weight with lay juries and judges. This arguably happened in *Wells v. Ortho Pharmaceutical Corp.*,[43] a products liability case alleging that a spermicide produced by Ortho had caused severe birth defects in a baby girl. Federal district judge Marvin Shoob awarded the plaintiff $4.6 million and her mother about $500,000, largely on the basis of his assessment that the scientists testifying for them were more believable than Ortho's witnesses. Dr. Bruce Buehler, for example, was commended for his steadiness on the stand, a quality that Shoob accepted as an indicator of good reasoning:

> His opinion at trial was the same as the opinion that he previously had offered at his deposition . . . His detailed explanation of how he had ruled out other possible causes demonstrated that his opinion was the product of a careful, methodical reasoning process, not mere speculation. His demeanor as a witness was excellent: he answered all questions fairly and openly in a balanced manner, translating technical terms and findings into common, understandable language, and he gave no hint of bias or prejudice.[44]

By contrast, the judge discounted the testimony of an important witness for Ortho because there was a striking "difference between his apparent certainty on direct examination and his less-than-certain tone on cross." Concerning another Ortho witness, Shoob concluded that his "criticisms of plaintiff's attorneys and of expert witnesses who testify for plaintiffs in malformation lawsuits," as well as "the absolute terms in which he expressed his conclusions," severely damaged his credibility.[45]

This case illustrates some striking differences between credibility determinations in law and science. For Judge Shoob, the testifying expert's personal credibility completely determined his scientific credibility. By contrast, the public ethos of science holds that the truth of a claim must be assessed independently of the personal attributes of the individual making that claim.[46] Scientists assert credibility by grounding their observations in bodies of communally accredited knowledge. During a legal trial, however, relatively equal numbers of experts are pitted against each other on the witness stand, giving the possibly false impression that opinion in the larger scientific community is also similarly divided.[47] Since both sides tend to seek the opinion that is most supportive of their positions, the adjudicatory process is confronted with two diametrically opposed opinions and cannot easily determine which is genuinely deviant and which (if any) is held by a substantial body of knowledgeable experts.

DNA Typing: A Shaky Consensus

Litigation becomes an effective vehicle for civic education when the adversary process succeeds in exposing the unacknowledged and untested presumptions concealed within a seemingly robust scientific claim, field, or technique. The legal system belatedly served just this function in a series of cases involving the use of DNA typing in the 1980s. Introduced into U.S. criminal proceedings in 1986, DNA test results were hailed at first as the answer to every law-enforcement officer's dreams. A widely used version of the technique produces a visual representation of repetitive DNA sequences of varying length ("variable number of tandem repeats," or VNTRs), which are statistically extremely unlikely to be the same in any two people. Like traditional fingerprints, a DNA "fingerprint" is said to provide virtually unassailable evidence of identity (or lack of identity) between a blood, semen, or other sample taken from a suspect and one taken from the scene of the crime. Within four years of its first use, evidence from DNA tests had been admitted in some 185 cases around the country.[48] Yet by the end of this period, this apparently unimpeachable form of testimony fell victim to arguments that revealed the subtly constructed foundation of its supposed infallibility.

The consensus about DNA typing began to fall apart in the landmark case of *People v. Castro*,[49] when experts for the defense and prosecution joined together to challenge the sloppy and as yet unstandardized procedures used by Lifecodes Corporation, a private testing laboratory, in interpreting DNA evidence. Subsequently, in *Maine v. McLeod*,[50] a sexual molestation case, defense experts again questioned the way in which Lifecodes had identified a match between two DNA samples. The two "fingerprints" in this case were not in fact identical to the untrained eye: although the pattern was the same, the bands in one print were displaced relative to the other in a way that suggested that the DNA fragments in the two samples were of different lengths. A Lifecodes expert testified at trial that the company had used an unvalidated methodology (that is, one not reviewed or approved by other experts) to correct for the observed bandshift. This admission fatally weakened the supposed identity determination.

In other trials, efforts to establish the defendant's identity beyond doubt were successfully challenged on the ground that the prosecution had not established the prevalence of particular DNA markers in relevant population groups. In early cases, experts had predicted extremely low probabilities of accidental matches by assuming that each sequence tested in DNA typing was independent of every other. This assumption overlooked the possibility that certain patterns of sequences might be found with greater frequency among members of the same ethnic group. Legal fact-finders had initially been unaware of the need for statistical data from the field of population genetics to back up experts' probability claims. In an effective display of boundary work, DNA fingerprinting was originally represented by its proponents as a taken-for-granted technique belonging only to the fields of molecular genetics and molecular biology.[51]

It was not cross-examination pure and simple that led lawyers and members of the scientific community to deconstruct the consensus that had passed unchallenged in the early DNA fingerprinting cases. Rather, the process of deconstruction was set in motion because the legal system's normative commitment to finding two sides in every case led in *Castro* to a confrontation among experts who finally questioned some of the methodological premises of DNA testing and testimony. But once Pandora's box was opened, scientists such as MIT's Eric Lander, who had enthusiastically participated in the box's initial opening, found to their dismay that it is not so easy to put the lid back on the process of deconstruction. A 1992 report by the National Research Council failed to settle the questions that had been raised by population geneticists. The NRC study committee proposed a method of calculating probabilities, known as the "ceiling principle," which would compensate for data gaps and yet be conservative enough to satisfy civil libertarians. Instead of resolving debate, the report sparked renewed controversy, and a second committee was appointed in 1994 to review the matter further.[52] The *Nature* article by Lander and Budowle was too clearly colored by the authors' interest in guaranteeing the acceptance of DNA typing to close the debate; indeed, it was challenged almost immediately in the letter pages of *Nature* by Richard Lewontin and Daniel Hartl, the two Harvard biologists most prominently associated with skeptical attacks on the statistical and methodological reliability of DNA typing technology.[53] These episodes lead us to question Washington attorney Milton Wessel's optimistic assertion that "for issues of sufficient societal significance, the 'state of the science conference' approach will produce a scientific consensus of such persuasive character that few will attack it even in the most adversarial legal proceeding."[54] Not so, it seems, for DNA typing.

Judicial Gatekeeping and the Reconstruction of Expertise

If adversary procedures are calculated to deconstruct expert testimony, then how does the legal system achieve closure on scientific issues, let alone sustain its image as a forum for arriving at the truth? For answers, we turn to the gatekeeping function that courts perform in relation to expert evidence, whether by directly screening experts or by deferring to real or imagined external sources of technical authority.

Certifying the Expert

A proposed witness's qualifications to testify in court are first established through a *voir dire* examination. On the basis of questions asked by counsel for both parties, the trial judge forms an initial opinion of the expert's claim to specialized knowledge and determines whether the witness should be admitted or not. The entry barrier at this stage is relatively low (for example, a witness may claim expert status on the basis of experience alone, without benefit of formal credentials), and the trial judge's decision to admit a witness is for all practical purposes final.

According to the Federal Rules of Evidence enacted by the U.S. Congress in 1975, the basic function of the qualified expert witness is to assist the legal fact-finder in understanding technical evidence. As a corollary, the rules affirm that judges may screen out irrelevant or inappropriate evidence. Thus, Rule 403 permits the court to exclude evidence whose "probative value is substantially outweighed by the danger of unfair prejudice, confusion of the issues, or misleading the jury." Under Rule 703, judges are free to ascertain that the data relied on by the expert are "of a type reasonably relied upon by experts in the field." In practice, judges differ widely in their willingness to exercise this legal authority. In a New York case, for example, a multimillion-dollar judgment turned on the claim that the chemical perchloroethylene had caused the plaintiff's kidney failure. At trial, a review of the extensive scientific literature on perchloroethylene failed to connect the chemical with kidney disease; the plaintiff's condition, moreover, was shown to be chronic and not likely to have been caused by a chemical agent. Nevertheless, an internist with no special knowledge of perchloroethylene or of chemically induced diseases was permitted to testify on the claimant's behalf, and the jury accepted his opinion on causation as the basis for a verdict in favor of the plaintiff.[55]

The Fifth Circuit Court of Appeals, by contrast, aggressively asserted its gatekeeping role in reversing a verdict for a railroad worker who claimed to have contracted porphyria through exposure to dioxin.[56] Two of the plaintiff's witnesses, a chemical engineer and a toxicologist, had testified that the plaintiff's disease could have been caused by dioxin, although neither claimed any personal knowledge of the extent of the plaintiff's exposure to the compound. Experts for the defendant, Monsanto, testified that the worker's maximum exposure must have been less than 2 percent of levels permitted in residential areas and was in any event insufficient to cause porphyria. The appellate court ruled that the evidence offered for the plaintiff was insufficient under these circumstances to support the verdict in his favor.

Activist judges, who use the powers granted by the Federal Rules of Evidence, necessarily inject their own perceptions about "what the science says" into the cause–effect picture that is constructed in the courtroom. Vietnam veterans who opted out of the settlement in the Agent Orange case objected to just such an intrusion by presiding judge Jack Weinstein, who had refused to let their claims proceed to trial. In a summary judgment dismissing their claims, Weinstein determined that the veterans had raised no scientific issues suitable for resolution by a jury. To reach this conclusion, the judge relied on his unaided assessment that neither the animal studies nor the epidemiological studies offered by the plaintiffs were trustworthy.[57] The plaintiffs' attorneys argued that Weinstein should have relied upon expert testimony in deciding whether these studies provided a reasonable foundation for a trial. The Second Circuit, however, dismissed their challenge, showing that federal courts jealously guard the right to make independent assessments of experts' reliability.[58]

More insidiously perhaps, judges are swayed by their perceptions of what "science" is and who is a "scientist" when they certify an expert's credibility. The boundary work by which courts hold experts to their own subjective standards of credibility became clear in a series of decisions about the admissibility of blood-typing data developed through a technique called electrophoresis. On the basis of testimony by the same key expert witnesses, the Supreme Court of Kansas admitted the evidence, whereas the Michigan Supreme Court held that it was not yet admissible. Different concepts of "science" and "scientists" guided each court's thinking. The court in Kansas saw all the witnesses as equally expert and hence as equally entitled to deference; the Michigan court, by contrast, characterized the main prosecution witnesses as "technicians" rather than as "scientists" and used this distinction to discount their claims about the technique's reliability.[59]

Quite unremarkable cases often provide telling glimpses into the judiciary's tacit understanding of what constitutes science. Thus, a California court disclosed its biases when ruling on the admissibility of the horizontal gaze nystagmus (HGN) test administered by police officers: "The observation of HGN in a person and its interpretation as an effect of alcohol intoxication do not necessarily require expertise in physiology, toxicology, or any other scientific field. The nystagmus effect can be observed without mechanical, electronic or chemical equipment of any kind . . . it requires no more medical training than administration of the other field sobriety tests, such as the one-legged balance."[60] The criteria that the court used here to distinguish scientific from nonscientific observations—the existence of a named discipline, the availability of specialized training, and the use of equipment—were not open to challenge unless they could be represented, on appeal, as an error of law. Yet they reflected nothing more than one judge's untutored appreciation of the differences between scientific and other forms of knowledge.

In *People v. Williams,* another California court ruled that a police officer could not offer evidence from the HGN test as a lay person, since he could testify only "because of his knowledge, training, and experience which was clearly beyond common experience." But neither could he testify as an expert: his testimony lacked "understanding of the processes by which alcohol ingestion produces nystagmus, how strong the correlation is, how other possible causes might be masked, what margin of error has been shown in statistical surveys, and a host of other relevant factors."[61] Again, the court's judgments about the boundary between "common experience" and "expertise" were not available for debate except within the formal constraints of a legal appeal. This opaqueness of judicial constructions of science and expertise is a fact to be reckoned with in a democratic society, even if one sympathizes with the *Williams* court's reluctance to let a police officer clothe his unmediated visual observations of a suspect's eyeball movements in the mantle of expertise.

Deferring to Scientists

While insisting on the right to determine what evidence is admissible, U.S. courts have experimented for much of the century with guidelines for deferring to experts in making such decisions. The notion that judges should bow to an expert consensus was first articulated in *Frye v. United States,* a 1923 murder trial in the federal Court of Appeals for the District of Columbia Circuit. A cryptic but much-quoted passage from that brief opinion became known as the *Frye* rule:

> Just when a scientific principle or discovery crosses the line between the experimental and demonstrable stages is difficult to define. Somewhere in this twilight zone the evidential force of the principle must be recognized, and while courts will go a long way in admitting expert testimony deduced from a well-recognized scientific principle or discovery, the thing from which the deduction is made must be sufficiently established to have gained general acceptance in the particular field in which it belongs.[62]

Frye itself employed the "general acceptance" standard to exclude a lie-detector test offered by the defense; Frye was convicted of murder but was subsequently released when new exculpatory evidence came to light.

For decades, the *Frye* rule generated sharp disagreements within the legal community on both philosophical and pragmatic grounds.[63] Some liked the idea of submitting the issue of reliability to the scientific community, which would act like a "technical jury," rather than to a trial judge with no particular qualifications to pass on the status of a new technique.[64] Others objected to the vagueness of the standard, which in their view promoted inconsistent decisionmaking. Although *Frye* posited that something less than universal agreement (namely, *general* agreement) would suffice for admitting scientific testimony, it did not provide guidance as to how much agreement was enough, or among whom. Courts accordingly diverged in the degree of consensus they deemed necessary for establishing general acceptance. Variations in boundary work, as in the electrophoresis cases noted above, also produced contradictory results. Judges around the country disagreed on whether radar detection devices could be used to establish speeding violations, voice-prints to prove a speaker's identity,[65] or the statistical analysis of literary style (stylometry) to establish the authorship of a written statement.[66] Inconsistencies in applying the *Frye* rule led some courts to repudiate it altogether in favor of the balancing approach prescribed by Rule 403. Upholding the use of controversial voice-print testimony in a narcotics trial, the U.S. Court of Appeals for the Second Circuit, in *United States v. Williams*,[67] reaffirmed the trial court's right to weigh the probative value of the evidence against its potential for misleading or confusing the jury. The court concluded that it was improper to "surrender to scientists the responsibility for determining the reliability" of evidence.

Seventy years after *Frye*, the U.S. Supreme Court finally agreed to decide whether the general acceptance test should continue to govern the admissibility of scientific evidence. Although the legal profession had by now accumulated a lifetime of wisdom on this issue, the *amicus curiae* (friends of the court) briefs submitted by various professional groups in *Daubert v. Merrell Dow Pharmaceuticals, Inc.*[68] showed little evidence of systematic learning from the problems of *Frye*. The recommendations before the Court fell roughly into two camps. One school, led by such weighty professional organizations as the American Association for the Advancement of Science (AAAS) and the National Academy of Sciences, urged the Court to look to science for sources of validation, in particular, by accepting peer review as the primary basis for distinguishing acceptable from unacceptable science. Another school took a more charitable view of the ability of judges to make the necessary discriminations and proposed tests that would import into the law what they saw as universal criteria for determining a claim's scientific status. The latter approach presumed that science, as a unique method of inquiry, could be distinguished from other ways of acquiring knowledge through formal criteria such as testability (see the boxed table on the next page for statements of the principal positions).

The majority opinion in *Daubert*, written by Justice Harry Blackmun, regarded at that time as the Court's leading authority on science, borrowed with remarkable eclecticism from all the representations before it. The petitioners won a formal victory as the Court strongly reaffirmed the discretionary power of judges to assess and screen scientific evidence; the Court also sided with the petitioners in rejecting both peer review and general acceptance as absolute markers of reliability. *Frye*, in this respect, was definitively supplanted by the approach taken in the Federal Rules of Evidence. But the respondents and the scientific community also won some comfort in that the Court exhorted judges to evaluate science in accordance with methods deemed acceptable by scientists. To operationalize this mandate, the majority opinion posited four criteria for determining the validity and relevance of proposed testimony: (1) whether the theory or technique underlying the evidence has been tested and is falsifiable; (2) whether it has been peer reviewed; (2) the technique's error rate, if known; and (4) general acceptance, which now reappeared as just one among several factors relevant to the issue of admissibility. Tellingly, the Court managed to weave into this rehearsal both the view of the philosopher Karl Popper that science progresses through clear falsifications of erroneous claims and the view of constructivist sociologists of science that knowledge accumulates through negotiation and consensus among members of the scientific community.[69]

Science and Law after *Daubert*

Daubert's fine disregard for a philosophically coherent decision rule on admissibility may exemplify the common law's genius for muddling through on the basis of experience rather than logic.[70] Its mixed message, however, invited others to clarify what Justice Blackmun had left ambiguous.

Daubert v. Merrell Dow Pharmaceuticals, Inc.: Excerpts from *Amicus* Briefs to the Supreme Court

American Association for the Advancement of Science and the National Academy of Sciences, in support of Respondents: "Courts should admit scientific evidence only if it reasonably conforms to scientific standards and is derived from methods that are generally accepted as valid and reliable. Such a test for admissibility would incorporate the factors, including the results of peer review, that scientists consider in reviewing each other's work."

Carnegie Commission on Science, Technology, and Government: "In order to make this inquiry [whether scientific claims have been developed within the bounds of a recognizable form of scientific inquiry], the court should consider the following criteria: (1) Is the claim being put forth testable? (2) Has the claim been empirically tested? (3) Has testing been carried out according to a scientific methodology?"

Petitioners: "But the ultimate determination of the persuasiveness of an expert's opinion has been entrusted by the Federal Rules to the jurors . . . not to the editors of private journals nor even [to] judges who wish to defer to such editors. The peer-review process has many virtues but . . . screening the reliability of experts' views for use in federal court is not among them, for neither the objectives of peer review nor its procedures are suited to the determination of truth."

Respondents: "For some scientific claims, the necessary foundation requires out-of-court passage through the process of scientific community scrutiny and verification; for other claims, the foundation requires conformity of the underlying reasoning to accepted scientific standards for validation of that type of claim."

Chubin et al., in support of Petitioners: "to suggest, as the Ninth Circuit did below, that the appearance of a paper in a peer review journal betokens 'general acceptance' by a 'consensus' of the scientific community is senseless. If anything, the numerous rebuffs that generally precede eventual acceptance connotes [sic] 'general rejection' much more than a single endorsement manifests 'general acceptance.'"

The initial response from the lower courts included some reassuring signs from the standpoint of industry and the defense bar. Upon reconsidering the *Daubert* evidence in the light of the Supreme Court opinion, the Ninth Circuit Court of Appeals concluded that the contested reanalysis of epidemiological data was not admissible.[71] Practicing attorneys as well as academic interpreters of *Daubert* tried in the meantime to read their own constructions of what science is back into the actual opinion. Thus, Ron Simon, a well-known plaintiff's lawyer, commended the Court for recognizing that "science is a very fluid process that allows for various simultaneous explanations of the same event."[72] By contrast, an article co-authored by Francisco Ayala, prominent scientist and AAAS president, and two attorneys complimented the Court on finally having directed the judiciary toward adopting wholesale scientists' own view of what counts as proper science. The image of science presented in this article appeared to be a product of the authors' introspection rather than research. Eschewing any serious engagement with contemporary social studies of science, the authors observed in a revealing footnote, "We make no claim to philosophical rigor, or to resolving the positivist versus relativist and other debates about the nature of science. Instead, our discussion aims to present a picture of science in accordance with *the way most scientists actually practice their profession*" (emphasis added).[73]

Institutional issues remained as unsettled after *Daubert* as epistemological ones. Would it be necessary or even desirable to create new judicial or quasi-judicial institutions dedicated to the resolution of factual disputes? The idea of a "science court," proposed by engineer Arthur Kantrowitz in 1967 and further developed by a presidential task force in the mid-1970s, has largely been abandoned as unworkable.[74] As originally conceived, the science court was to serve as a vehicle for separating facts from policy and for addressing factual issues competently. Advocates saw the court as a safeguard against "efforts to impose the value systems of scientific advisers on other people" and to prevent "the extension of authority beyond competence."[75] With time, critics concluded that many supposedly factual disputes would only serve as surrogates for more basic disagreements about the social or moral viability of particular technological choices.[76] These discussions were quite consistent with the gradually increasing awareness of scientific knowledge as a special type of social construct.

A more modest approach to institutional reform would build on the idea of strengthening the technical supports available to judges. Frequently discussed options include the increased use of court-appointed experts, special masters or technically trained law clerks in complex cases. A formal means of moving in this direction is already available under the Federal Rules of Evidence, which grant judges broad powers to seek help from court-appointed experts or panels if they believe that such procedures will assist the process of scientific fact-finding. This power, in practice, is only rarely used by the federal courts. A study carried out by the Federal Judicial Center in 1989 found that only 19 percent of federal judges used the authority granted under Rule 706, although 87 percent indicated that such authority would be helpful at times, especially in patent and products liability cases.[77]

Support for the greater use of court-appointed experts rests on the largely unexamined assumption that the removal of fact-making from the exclusive control of the disputing parties will produce an unbiased account of scientific evidence. According to John Langbein of Yale Law School, the great advantage of this approach would be to convert the parties from "law-and-fact adversaries" to mostly "law-adversaries" on the model of Continental litigators.[78] Langbein, who implicitly accepts the possibility of neutral fact-finding, points to European experience to suggest that the results could reduce waste and promote impartiality. Efficiency in the European mode, however, could be bought only at the price of privileging one set of biases over another.[79] Brian Wynne's study of the British public inquiry into the construction of the Windscale nuclear fuels reprocessing plant shows how, in a relatively closed expert culture, judicial fact-finding was used to shore up the dominant sources of scientific and institutional authority.[80]

Proposals concerning law clerks and special masters have also proved problematic in the light of experience. In the early 1950s, for example, Carl Kaysen, a distinguished economist, served as law clerk to Judge Wyzanski of the Massachusetts federal district court in a landmark antitrust case.[81] The opinion he helped author won wide praise and was affirmed by the Supreme Court. But Wyzanski himself eventually concluded that his reliance on Kaysen had been unwise: no judge could maintain the "requisite independence when the assistant had, in effect, 'mastery' over the judge through superior technical ability, learning and experience."[82] To alleviate some of these concerns, Wyzanski in another case made his relations with an expert assistant more public; the special master conferred with each of the parties in the other's presence, submitted a report, and was available for examination in open court. Within the American legal tradition, such efforts to enhance the educational and mediating functions of court-appointed experts may well find quicker acceptance than the notion that judges should have access to expertise superior to that of the disputing parties.

There are practical difficulties as well in using scientifically trained law clerks in the ordinary courts. In most federal jurisdictions, chronic overloading of the docket requires that clerks be available for rotation from one case to another as need arises. Given the hectic and unpredictable pace of decision-making, few judges are prepared to hire law clerks with the guarantee that they would be assigned only to cases relating to their scientific specialty. Accordingly, it may be impossible to get a perfect match between clerk and case unless a lawsuit is complex enough to demand the full attention of a suitably trained specialist law clerk.

Looking Ahead

Science is constructed in the courtroom in accordance with tightly circumscribed rhetorical and procedural rules, under unavoidable economic and sociological constraints, and to serve widely divergent normative agendas. Altering any component of the current system for using expert evidence would change the balance of interests that have become engaged in constructing and deconstructing scientific facts in the legal arena. New procedures would not free science from moral and social pressures or enable scientists to offer pure factual guidance to the law, as the technocratic ideal persistently imagines. Rather, the production of scientific facts would simply be embedded in a different matrix of contingencies: thus, the power of judges might be enhanced at the expense of the litigants, and new interests, such as universities or professional societies, might gain a stake in the proceedings through changes in the rules for screening expert witnesses.

A less politically contentious strategy would be to educate judges, lawyers, and scientific experts in each other's modes of reasoning and discourse. Given the frequency of toxic tort actions, for example, it may be reasonable for federal and state judges to acquire a basic familiarity with the principles of epidemiological studies, cancer bioassays, and risk analysis; the Federal Judicial Center embarked on such a project to increase the scientific literacy of judges with its 1994 *Reference Manual on Scientific Evidence*.[83] The experience of Ortho's witnesses in *Wells* indicates that experts, for their part, might also do well to familiarize themselves with the aims of the legal process and the standards that legal fact-finders use in assessing truthfulness. Another approach to reform would be to resolve controversies by means that do not necessitate the capture of science in the service of adversarial interests. Lowering the stakes for both winners and losers, for instance through expanded, no-fault insurance systems coupled with caps on damages (especially punitive damages) and fees, might serve justice while reducing the temptation for purely opportunistic uses of science. We will return to these points in Chapter 10, following a more fine-grained analysis of judicial treatments of science and technology in specific decisionmaking contexts.

4

The Technical Discourse
of Government

The American legal system's power to subordinate technical fact-finding to normative policy goals emerges with special clarity in the context of judicial review, where courts enjoy an authority unparalleled in the Western world to question the expert judgments of executive agencies. It is a basic tenet of U.S. administrative law that agencies must use their delegated powers rationally, including, of course, the power to interpret uncertain or ambiguous science. Under law, moreover, courts are the ultimate supervisors of administrative rationality. Thus, the Administrative Procedure Act of 1946 provides that courts may invalidate agency decisions if they are not based on "substantial evidence" or if they are "arbitrary, capricious, an abuse of discretion, or otherwise not in accordance with law." Similarly, the major regulatory statutes of the 1970s contain judicial review provisions authorizing the courts to strike down agency decisions unless there is a reasonable connection between the proposed action and the record that supports it. Paradoxically, therefore, lay judges have the power to overturn decisions made by administrative agencies with considerably greater technical expertise and policy experience.

The gap in expertise between courts and agencies has made it difficult for courts to decide how energetically they should exercise their review function. On the one hand, judges recognize that they cannot lightly second-guess a regulatory agency's informed opinion on matters within its competence. On the other hand, it is equally apparent that some reappraisal of the technical record is needed in order to make judicial review meaningful. Balancing these considerations is particularly challenging when courts are asked to review the complex rulemaking records characteristic of modern health, safety, and environmental regulation. It took two decades of wavering effort to sort through this puzzle and to establish something approaching a coherent boundary between agency autonomy and the supervisory reach of the courts. Even then, the judicially crafted boundary rested on an artificially optimistic assessment of the agencies' technical capacity and political legitimacy; ironically, it appears that the scientific community may recently have seized from courts the initiative for defining decisionmaking procedures that correspond better with agency capacity.

Looking back over this period, we can discern three distinct phases in the articulation of legal norms for the review of science-based regulatory decisions. In the first phase, which lasted through the 1970s, legal doctrines concerning the scope of judicial review developed in response to challenges to federal health, safety, and environmental regulation. Novel review petitions flooded the courts under regulatory statutes that considerably expanded the power of the judiciary to hold agencies accountable for their actions. Courts demonstrated a willingness to probe the scientific underpinnings of administrative actions and to demand reasoned explanations for agency interpretations of controversial data. Cases in which administrators had failed to act on the basis of new evidence attracted especially searching attention.

The 1980s ushered in a second phase, marked by growing skepticism toward the use of risk assessment and other predictive methodologies for regulating technological hazards. Courts worried that the doctrines developed during the previous decade did not sufficiently constrain agencies in interpreting scientific uncertainty, and they tried with mixed success to impose more stringent limits on administrative discretion.

By the mid-1980s, however, judicial activism in reviewing technical decisions began to wane. The swing back to a more deferential style of review in this third phase reduced the political saliency of judicial review in the short run, but it left unresolved important questions about how to conduct the technical discourse of government in a complex, modern democracy.

The historical record suggests that the way judges defined their mission in reviewing science-based administrative decisions was strongly influenced by broader trends in national politics. Thus, when the activist, pro-environmental attitudes of the 1970s yielded to a more probusiness and antiregulatory politics in the 1980s, the philosophy of review followed suit. Yet although they were responsive to the pressures of national politics, courts did not completely back away from the demand for public accountability in regulatory decisionmaking; nor did they become, even in the conservative Reagan-Bush era, mere mouthpieces for White House policy. Two decades of judicial review helped shape a regulatory culture in which administrators are expected to explain their scientific assessments just as explicitly as their interpretations of the law. The ruling that required First Lady Hillary Rodham Clinton to open up her health care task force to the public was a reminder that expertise in governmental policymaking must remain accountable to democratic values.[1]

The Rise of Precautionary Regulation

We saw in Chapter 2 that a significant shift in regulatory thinking occurred during the late 1960s as Congress and the public jointly propelled federal agencies into the business of preventive policymaking. Instead of regulating technologies that were known to be harmful, agencies now were charged with identifying and controlling hazards that could, if unchecked, pose serious threats to health, safety, and the environment. Risk (that is, the possibility of harm), rather than harm itself, became the phenomenon that Congress asked the agencies to guard against, particularly when such risks were determined to be "unreasonable."

The move from a harm-based to a risk-based standard for regulatory action greatly expanded the discretionary powers of the agencies. To justify intervention, administrators no longer needed visible evidence of dying workers, decaying forests, or rivers on fire with combustible pollution. The mandate to assess risks at once expanded the categories of evidence that administrators could view as relevant and the range of activities that they could characterize as dangerous. In risk-based regulation, there was by definition no direct proof of human disease or death. The agencies increasingly took their regulatory signals from studies involving laboratory animals, and attempts to extrapolate human risk estimates from animal data developed into a persistent source of controversy. Scientific uncertainty, marked by open disagreement among experts, emerged as the most serious impediment to effective environmental and public health decisionmaking.

As federal regulation began impinging on a wider cross-section of hazardous activities—chemical and pharmaceutical manufacture, petroleum refining, and electric power generation, to name but a few—industry leaders recognized that the risk assessments upon which agencies were building their programs of preventive regulation presented an easy target. Based on untried methods of data collection and novel, often cross-disciplinary interpretive judgments, agency risk assessments lacked the institutional credibility of normal science. New mission-conscious agencies, such as the Environmental Protection Agency (EPA), the Occupational Safety and Health Administration (OSHA), and the Consumer Product Safety Commission (CPSC), soon found themselves at loggerheads with industry over the interpretation of scientific evidence, especially when evaluating the risk of cancer from exposure to toxic substances. While industry complained of faulty science, a new generation of public interest activists, concerned that congressional mandates to protect the environment were not being vigorously implemented, argued that the agencies were moving too slowly to relax their traditional reliance on proof of harm. Seeking to effectuate their particular philosophies of regulation, critics of both camps took the agencies to court for alleged flaws and failings in their use of science.

Administrative Accountability
and Scientific Conflict

U.S. regulators are particularly vulnerable to charges that they have misused science.[2] The growth of social regulation since the 1970s has expanded the discretionary powers of regulatory agencies but has also pressured them to elaborate the technical rationale for controversial policy choices in extensive and painstaking detail.[3] The rulemaking record (that is, the government's official account of its actions) must spell out the agency's scientific assumptions, uncertainties, and judgments, as well as those matters that the agency accepts as valid knowledge.[4] Competing interpretations of the evidence must be represented along with the reasons for rejecting them. Opinions that support the agency's final judgment must be lined up against countervailing arguments, also noted in the record, that could have led to substantially different regulatory conclusions. Structured by these rules, the record becomes a history more of conflict than of consensus.

Attempts to regulate carcinogenic chemicals since the 1970s document the fragility of this highly public approach to securing scientific legitimacy.[5] Chemical carcinogens were among the first pollutants that activist groups targeted for stricter regulation at the dawn of the modern environmental movement. Relying on the precautionary mandates of statutes such as the Federal Insecticide, Fungicide, and Rodenticide Act (FIFRA), the Occupational Safety and Health Act, and the Delaney Clause of the Federal Food, Drug, and Cosmetic Act, environmental, labor, and consumer organizations urged federal regulators to ban substances that had been shown to cause cancer in animals, whether or not there was proof of harm to human beings. Chemical manufacturers vehemently resisted this effort, fearful that expanded use of animal evidence would increase the costs of product development and drive a substantial percentage of their most profitable products off the market.

In the ensuing legal confrontations, both sides mustered substantial scientific as well as policy support for their positions. Environmentalists gained critical leverage from the ob-

servation that virtually all known human carcinogens also
caused cancer in animals. Moreover, since ethical considerations precluded studies of human populations, animal data
provided the best experimental evidence of the probable effects
of chemicals on human health. Numerous methodological
choices that increased the chances of producing false positives
(or exaggeratedly high estimates of risk) in animal tests won
support on pragmatic grounds. For example, proregulation
groups strongly supported the practice of exposing animals to
high doses of suspected carcinogens, since this was the only
practical way of detecting relatively low risks while working
with manageably small populations of test animals. It became
standard, if always contested, practice to conduct studies at
the "maximum tolerated dose" (MTD), the highest dose at
which the animals would survive the study period without suffering severe toxic effects.

Industry scientists produced a barrage of counterarguments
about the wisdom of using animal studies alone as a basis for
assessing risk to humans. Focusing on the biological differences between animals and humans, chemical manufacturers
argued that the positive carcinogenic response seen in many
animal studies was specific to the tested species and hence
should not be generalized to humans. High-dose testing was
another issue that provoked endless controversy because of its
power to produce false positives. Industry experts alleged that
exposure to the MTD frequently induced metabolic and other
changes in the tested animals that heightened their susceptibility to cancer. Such changes would not occur at the lower
exposures actually experienced by human populations. Finally, the assumptions made in extrapolating risk estimates
from high-dose animal tests to humans at much lower exposure levels caused increasingly bitter debate. Scientists from
industry attacked key elements in the risk assessment techniques developed by EPA and other agencies as overly conservative and lacking scientific support. The use of mathematical
models to derive quantitative estimates of the risk of cancer
attracted particularly strong opposition on the grounds that
they bore little relation to biological reality and masked the

poor quality of experimental data with misleadingly precise calculations of risk.

As scientifically grounded challenges to carcinogen regulation grew more frequent, courts were placed in the uncomfortable position of having to evaluate the relative cogency of the interpretations advanced by the agencies and their detractors. The earliest cases required the courts to chart new territory in administrative law. They had to define, as an initial matter, how intensively they should scrutinize administrative decisions based on complex and uncertain science. Secondarily, they had to develop principled standards for controlling the discretionary judgments made by regulatory agencies under conditions of scientific and legal uncertainty.

The "Hard Look" Doctrine

What should a reviewing court seek to achieve in supervising technical decisionmaking? Holding an expert agency accountable to legal requirements and democratic values is an important part of the task, but to do so without second-guessing the agency's informed judgments requires courts to strike a delicate balance. The judiciary has always concurred wholeheartedly with Judge David Bazelon's observation that "substantive review of mathematical and scientific evidence by technically illiterate judges is dangerously unreliable."[6] Even when an agency's "expertise" consists chiefly of familiarity with the mechanics of a particular regulatory program, it is difficult for the reviewing court, an institutional outsider, to decide whether the agency has applied its expert knowledge legitimately to the problem at hand. Yet courts remain a reasonably effective institutional key for opening up a bureaucratic-technical culture that operates, for the most part, out of the public eye. Doctrines of judicial review represent a struggle to define just how much transparency to bring to the normally unseen workings of regulation.

Early in the era of social regulation,[7] courts announced their intention to enter into an active partnership with the agencies. Judges would not supplant agency decisionmakers in making policy but would hold them to a high standard of technical accountability by making sure that they had taken a "hard look" at the evidence before them. The primary duty of the courts would be to ensure that the agency's decision was "reasoned," and in opinion after opinion reviewing judges indicated that they would not treat the notion of reasoned decisionmaking lightly. To satisfy their judicial critics, agencies would have to meet stringent criteria of clarity and consistency: "Assumptions must be spelled out, inconsistencies explained, methodologies disclosed, contradictory evidence rebutted, record references solidly grounded, guesswork eliminated and conclusions supported 'in a manner capable of judicial understanding.'"[8]

The "hard look" doctrine seemed on the surface to provide the courts with a plausible formula for balancing their review function against the primary decisionmaking powers of technically expert agencies. As a practical matter, however, the "hard look" failed to establish a sustainable operational boundary between rulemaking and judging. A split of opinion between Judges David Bazelon and Harold Leventhal, two influential members of the Court of Appeals for the District of Columbia Circuit, highlighted the difficulty of translating the standard into meaningful practice. Leventhal believed that judges would have to penetrate the substantive underpinnings of challenged decisions in order to determine whether the responsible agency had exercised reasoned discretion. He therefore thought it proper for judges to "steep" themselves in the technical issues before the agency and to "acquire whatever technical knowledge is necessary as background for decision of the legal questions."[9] As interpreted by Leventhal, the "hard look" metaphor subtly became a rationale for permitting reviewing judges themselves to look hard at the scientific arguments supporting agency decisions. Put differently, judges unavoidably became parties to the construction of the scientific case from which policy choices would ultimately follow.

Judge Bazelon was far less optimistic about the judiciary's capacity to master scientific intricacies. His view was that judges should focus on the procedural rather than the substantive aspects of rulemaking: by carefully monitoring an agency's procedural choices, courts could ensure that all relevant issues had been thoroughly aired and that experts holding different viewpoints had been given an opportunity to participate. If agencies fell short of meeting these standards, Bazelon proposed that courts should require administrators to adopt additional procedures (such as trial-type hearings accompanied by cross-examination) to ensure fuller elaboration of the relevant scientific arguments.

Following Bazelon's prescription, the D.C. Circuit on several occasions remanded agency decisions for failure to adopt sufficiently deliberative rulemaking procedures. This practice had the merit of requiring agencies to rethink their technical assessments case by case, contingent on the particular legal and scientific issues presented in a given context. Yet in the end activist procedural review failed to satisfy critics. First, Bazelon's approach did not really provide a guarantee of judicial restraint, since judges could not determine whether additional procedures were necessary without undertaking the searching substantive review recommended by Leventhal. Second, since knowledge relevant to regulation was constructed from a mix of facts, projections, and policy judgments, there were no externally given boundary markers that could warn judges when they were straying from mere procedural inquiry into the forbidden territories of the agency's scientific or experiential assessments. For these reasons, in *Vermont Yankee Nuclear Power Corp. v. Natural Resources Defense Council* a unanimous Supreme Court rejected court-prescribed procedures as misguided and unsound. Such "Monday morning quarterbacking,"[10] the Court opined, could only exacerbate the uncertainties of an already overburdened administrative process. And in any event, courts had no authority to impose procedures on the agencies beyond those explicitly required by Congress or by the due process requirements of the Constitution.

At the same time, *Vermont Yankee* affirmed that judges could properly scrutinize the substantive record on which decisions were based and ask agencies to reconsider weaknesses in the record. Leventhal's version of the "hard look" doctrine therefore retained currency through the 1970s, producing both positive and negative impacts on agency behavior. Inexperienced regulators frequently discovered on remand that they had indeed failed to consider relevant evidence or to take account of well-founded methodological criticism. The awareness that courts would engage in detailed factual review also prompted greater care in the construction of the decisionmaking record and the presentation of technical arguments.[11] These benefits, however, were won at the cost of overanalysis and growing regulatory indecision.

The Science Policy Paradigm

While federal courts, spearheaded by the D.C. Circuit, asserted their right to look hard at science-based decisions, they in practice granted the agencies considerable leeway in analyzing issues "on the frontiers of scientific knowledge." Asserting that such decisionmaking must "depend to a greater extent upon policy judgments than upon purely factual analysis,"[12] courts articulated a set of assumptions that reinforced the agencies' authority to provide the controlling interpretations of contested scientific data. Three principles emerged as central to this "science policy paradigm":[13] first, that regulatory decisions could be made on the basis of suggestive rather than conclusive evidence of harm; second, that interpretations would be deemed valid even if not backed by an expert consensus; and third, that it was within the administrator's discretion to select among disputed data or methodologies when confronted by expert disagreement.

These principles were influentially set forth in *Ethyl Corp. v. EPA*, which, as we saw in Chapter 2, represented the first serious attempt by a federal court to construe a precautionary standard for environmental protection. Judge J. Skelly Wright, the opinion's author, observed that the EPA administrator could not reasonably be expected to provide a "rigorous, step-by-step proof of cause and effect" when implementing a risk standard. Given the gaps in the evidence, it was more appropriate for the administrator to employ a mixture of factual analysis and expert judgment "to draw conclusions from suspected, but not completely substantiated, relationships between facts, from trends among facts, from theoretical projections from imperfect data, from probative preliminary data not yet certifiable as 'fact,' and the like."[14] This language strongly implied that agencies could rely on evidence that the scientific community was not yet prepared to regard as conclusive, and even on information that scientists viewed as technically inadequate. As a corollary, *Ethyl* confirmed the right of administrative agencies to determine for themselves whether evidence related to risk was substantial enough to support restrictive action.

Additional cases decided by the D.C. Circuit and other federal courts during the 1970s reaffirmed the elements of the science policy paradigm. For example, the courts upheld agency efforts to regulate chemical products on the basis of suggestive, but not conclusive, animal data indicating that they might cause cancer in humans.[15] In some of these cases, the courts explicitly stated that when dealing with a health threat so "sensitive and fright-laden"[16] as cancer, it was appropriate to regulate on the basis of "lower standards of proof than otherwise applicable."[17] Indeed, when regulators were attempting to guard against a public health hazard, the courts conceded that they might be held to less elaborate fact-finding procedures than in other contexts.[18] Finally, the courts indicated that even subclinical physiological effects (that is, changes not necessarily signaling the onset of disease) could be used as a valid basis for precautionary standard-setting.[19]

These decisions recognized that the adequacy of knowledge relevant to public decisions would have to be measured against the purposes for which knowledge was needed. Relatively less certain knowledge could justify taking action against extreme risks; "scientific validity" (like "general acceptance" under the *Frye* rule) was too blunt a standard to allow for the desired fine-tuning of the fit among risk, evidence, and action. But the science policy paradigm laid the groundwork for two problems that eventually undermined administrative credibility and confronted the courts with the need for additional controls on agency discretion. One source of tension was the courts' injunction that agencies did not have to apply "scientific" standards of proof in their assessment of evidence relating to risk. By granting administrators the freedom to construct less than "scientific" standards for assessing the validity of risk information, courts set the stage for future conflicts between the government and the scientific community. Each time administrative agencies chose to act upon inconclusive evidence of risk, they opened themselves to charges that they were relying on bad science rather than good policy judgment. Opponents of regulation had in effect been handed a basis for delegitimating the regulatory science done by agencies by comparing it with the "real science" done by scientists. The perception that agencies were acting on information unacceptable to members of the scientific community eroded the credibility of regulatory decisions even when they satisfied judicially created tests of legitimacy.

A second weakness of the science policy paradigm was that it left the agencies with what looked like too much discretion in standard-setting. Uncertainties in the knowledge base generally kept standards from being set at a fixed point on which all could agree. Although courts were acutely conscious of this fact, they acted as if statutory guidelines, supplemented by expert judgment, gave the agencies sufficient guidance about where to position a standard within the range of uncertainty. In practice, however, such guidance was reliably available only in the minority of cases in which a scientific consensus constrained agency discretion within fairly narrow limits.

In his study of the troubled enforcement history of the Clean Air Act, R. Shep Melnick argued that courts fundamentally misjudged the extent to which science could delimit standard-setting. Courts, in Melnick's view, wrongly assumed that EPA could find discernible thresholds for adverse health effects caused by air pollutants. Whether courts truly misunderstood the state of affairs or whether their commitment to an orderly administrative process led them to back up the agency's viewpoint is not entirely obvious. There is no doubt, however, that the assumption of scientific consensus—contradicted by the reality of controversy over health effects—allowed the courts to countenance a "health only" (that is, the regulator need only look at health benefits, not at costs) interpretation of key standard-setting provisions of the Clean Air Act during the 1970s. Consequently, the courts did not feel compelled to require EPA to carry out cost-benefit analyses in support of ambient air quality standards. Melnick concluded that this strategy "both weakened the bureaucratic position of the agency's economists and insulated it from pressure to relax proposed standards."[20]

Judicial review of regulatory decisions involving carcinogens similarly pushed agencies to err on the side of protection. Courts, for example, played a prominent part in ratifying, and even expanding, the congressional policy of giving higher priority to the regulation of carcinogens than to other environmental or health hazards. In *EDF v. Ruckelshaus,* a concerned D.C. Circuit extended the anticancer philosophy of the Delaney Clause from the area of food additives to EPA's regulatory program for pesticides.[21] The courts also established a double standard of review with respect to carcinogens, indicating that decisions not to regulate these substances would be scrutinized more carefully than decisions to take positive regulatory action.[22]

In the early years of social regulation, these trends in judicial decisionmaking clarified ambiguous statutory mandates, assisted hesitant new regulatory agencies in developing a sense of mission, and helped counteract bureaucratic and political pressures to adhere to reactive and harm-based approaches to regulation.[23] By 1980, however, the political climate for regulation had changed drastically. President Reagan was elected with an explicit mandate to deregulate and to make federal policy more sensitive to economic costs and consequences. In a suddenly more conservative environment, judicial support for aggressive regulation began to seem outmoded, and the courts felt called upon to cut back on the level of discretion enjoyed by administrative agencies within the framework of the science policy paradigm.

Curtailing Discretion in Risk Assessment

The second decade of judicial review of health, safety, and environmental decisionmaking began with a search for meaningful limits on regulatory agencies' scientific judgments. An important and largely uncontroversial doctrine to emerge from this period was that the mere existence of scientific uncertainty did not justify the regulation of trivial risks. The D.C. Circuit clearly conveyed this message to the Food and Drug Administration (FDA) in a decision concerning that agency's attempt to ban plastic beverage containers made of acrylonitrile polymer.[24] FDA had determined by means of a theoretical projection that minute (that is, below the level of detection) quantities of acrylonitrile, an animal carcinogen, might migrate from the container walls into the beverage. FDA therefore proposed to regulate acrylonitrile as a food additive subject to the Delaney clause, which banned the addition of carcinogens to food. Judge Leventhal, author of the "hard look" doctrine, held that to do so would amount to an abuse of discretion. No agency, he indicated, should feel compelled to regulate a risk of *de minimis* (insignificant) proportions merely because the governing statute conferred literal authority to make such a decision. In this case, the FDA commissioner was not obliged to apply the Delaney clause simply because in theory "the statutory net might sweep within the term 'food additive' a single molecule of any substance that finds its way into food."[25]

Other cases involving risk regulation in the 1980s, however, showed a continuing willingness on the part of the courts to investigate the scientific basis of agency decisions. Indeed, having ruled out procedural questions—was there adequate cross-examination, were all relevant parties heard—*Vermont Yankee* left open only the option of substantive review. The results were not always salutary. Even critics of the 1970s style of judicial activism admitted that there were remarkably few occasions when judges had actually misread scientific evidence.[26] By contrast, the Supreme Court's "benzene decision,"[27] possibly the most influential science policy case of the 1980s, was widely criticized as reflecting a naive judicial understanding of science in the regulatory process.

At issue in that case was a proposal by the Occupational Safety and Health Administration to reduce the workplace exposure level for benzene from 10 ppm (parts per million) to 1 ppm. OSHA argued that it was impossible in the current state of scientific knowledge to find a "safe" level of exposure to benzene, a known human carcinogen. Accordingly, the agency proposed to set the exposure standard at the lowest feasible level, as prescribed by its governing statute.[28]

The expected legal challenge from the petroleum industry brought the issue before the Supreme Court, where a plurality held that OSHA had omitted a crucial step in its regulatory analysis. Before regulating any chemical hazard, the plurality concluded, the secretary of labor had to satisfy a threshold legal requirement: to demonstrate that the risk at the existing standard was "significant" and that a new standard would bring about a measurable improvement to worker health.[29] The justices further proposed that, to establish significance, OSHA should carry out a quantitative risk assessment showing a numerical probability of harm. These prescriptions grew out of a legitimate concern that (as in the acrylonitrile case) governmental power should not be indiscriminately used to regulate low-level risks. The decision can be seen in this respect as a "temporary correction of political failure,"[30] with the Court supplying a rule of reason that was missing from both the statute and the agency's interpretation of it.

Yet the Court's own corrective mandate for OSHA was also flawed in that it displayed "a reflexive, unthinking reliance on mathematical assessments."[31] In striving to promote a commonsense view of what risks are reasonably worth regulating, the plurality revealed a lack of familiarity with essential principles of risk assessment, for example, that risk is a function of exposure as well as probability of harm.[32] More generally, the plurality appeared unaware of the "primitive state of the risk assessment art," a factor that allowed estimates differing by several orders of magnitude to be derived from the same underlying scientific data.[33] Arguably, a similarly misguided faith in the precision and predictive power of environmental assessments led the Supreme Court, in *Dolan v. Tigard,*[34] to ask municipalities for more detailed showings of "rough proportionality" in takings cases (see Chapter 2).

Another controversial judicial intrusion into risk assessment occurred in *Gulf South Insulation Co. v. Consumer Products Safety Commission.*[35] This case involved a challenge by the formaldehyde industry to the Consumer Product Safety Commission's proposal to ban a common home insulant, urea-formaldehyde foam insulation (UFFI). CPSC had determined that UFFI presented an unreasonable risk of cancer in part on the basis of a quantitative risk assessment that showed up to 51 additional cancer cases for every million persons exposed to the substance. Asked to review the decision, the U.S. Court of Appeals for the Fifth Circuit undertook a classic "hard look" into the facts and theories underlying CPSC's risk assessment and concluded that the agency lacked "substantial evidence" for its action. The court's aggressive reanalysis of CPSC's expert findings troubled many observers not only because it reversed the normal relationship between the agency and the reviewing court, but also because it revealed an inadequate judicial grasp of the relevant scientific principles. In particular, the court misunderstood the nature and purpose of animal tests and the statistical basis for high-dose to low-dose extrapolations.[36]

Decisions like these prompted legal scholars to reconsider the advisability of having judges actively reexamine scientific issues committed to the agencies, in short, to rethink the "hard look." There was in some quarters an uneasy feeling that "Judge Leventhal's call for scientific sophistication from 'generalist judges,' however admirable in theory, is unrealistic."[37] Still more disturbing was the realization that intensive substantive review of agency decisions did not necessarily provide meaningful guidance for future action and might in fact introduce additional uncertainties into policymaking based on risk assessment.[38] Thomas McGarity, a well-known administrative law expert, commented that the "hard look" metaphor had outlived its usefulness and that courts should seek a more restrained formula to circumscribe their review function.[39]

The Return to Deference

Problems in implementing the "hard look" doctrine account for a return by the mid-1980s to judicial deference in reviewing science-based regulation. The Supreme Court signaled its own commitment to judicial restraint most explicitly in *Baltimore Gas and Electric Co. v. Natural Resources Defense Council*

(NRDC).[40] The capstone of a long line of citizen suits against nuclear power, the case involved a challenge to the Nuclear Regulatory Commission's generic determination that permanent storage of high-level nuclear wastes would have no significant impact on the environment (the "zero-release assumption"). NRDC, the petitioning citizen group, questioned whether the Commission had adequately evaluated the uncertainties of long-term storage. As in earlier decisions on nuclear power,[41] the Supreme Court rejected the environmentalists' position, offering judicial restraint as its primary reason. The Commission, the Court observed, was "making predictions, within its area of special expertise, at the frontiers of science. When examining this kind of scientific determination, as opposed to simple findings of fact, a reviewing court must generally be at its most deferential."[42]

Baltimore Gas can reasonably be seen as a spiritual descendant of *Vermont Yankee* in that it extended to an agency's substantive decisions the same deference that the Court had earlier shown on procedural matters. Indeed, the decision can only be understood as a generalized plea for judicial restraint, since the Court failed to provide cogent arguments for why decisions "at the frontiers of science," where normative questions might intrinsically be expected to predominate, should be entitled to great respect from judges. The Court was seemingly untroubled by the Commission's remarkably reductionist assessment, which combined disparate, methodologically incommensurable risks from physical and man-made causes into a single point estimate of "zero release." "[T]he sheer volume of proceedings before the Commission," the Court approvingly observed, had adequately elucidated "the major uncertainty of long-term storage in bedded-salt repositories, which is that water could infiltrate the repository as a result of such diverse factors as geologic faulting, a meteor strike, or accidental or deliberate intrusion by man. The Commission noted that the probability of intrusion was small, and that the plasticity of salt would tend to heal some types of intrusions."[43] That these conclusions could have been deconstructed by an energetic opponent follows without question from everything we know about scientific controversy and the sociology of knowledge.

That the Court allowed them to stand unchallenged no doubt reflects its self-confessed awareness of the nation's reliance on nuclear power and its participation, along with the Nuclear Regulatory Commission, in a shared culture of technological optimism.

In *Chevron U.S.A. Inc. v. Natural Resources Defense Council*,[44] the Supreme Court went even further, holding that an administrative agency's interpretation of its legal mandate was also entitled to deference from reviewing courts. The case grew out of EPA's decision to apply the so-called bubble concept in regulating "stationary sources" under the Clean Air Act. EPA had originally construed the Act so as to treat each component of a plant as a separate source requiring a separate permit for increased emissions. Later, under pressure to adopt a more flexible policy, EPA amended its regulations so as to treat all emission sources within a plant as a single source encased within a hypothetical bubble. Emission increases anywhere within the bubble could now be offset against reductions elsewhere in the bubble without need for special dispensation from EPA. The Supreme Court accepted this approach as consistent with the law.

An interesting study of court decisions in the immediate aftermath of *Chevron* by Peter Schuck and E. Donald Elliott of Yale Law School suggests that the decision caused a significant drop in judicial remands for errors in substantive law.[45] Notably, though, the D.C. Circuit, which reviews the bulk of highly technical health, safety, and environmental regulations, actually increased its remand rate in the year after *Chevron;* even in 1988 the D.C. Circuit's affirmance rate was 14 percent below that in other circuits. Noteworthy as well are the findings that remands from the D.C. Circuit were less likely to make agencies drop their proposed actions than in other jurisdictions, and that "major changes" were less common following remands in the cluster of health, safety, and environmental regulation than in other regulatory areas. All this suggests that the highly formal, almost ritualistic, politics of risk regulation that developed in the 1980s under the D.C. Circuit's jurisdiction could not readily be subjected to external control, even by the Supreme Court.

Contextual Changes: Technocratic Judges and Democratic Experts

Seen against the backdrop of wider developments in the twentieth century, the trilogy of *Vermont Yankee, Baltimore Gas,* and *Chevron* looks strangely reactionary. By committing all borderline interpretive choices—whether on procedure or technical content or substantive law—to the discretion of the agencies, the Supreme Court tacitly espoused a model of the regulatory process that was deeply at odds with political reality. The Court's policy of deference was predicated on the capacity of regulatory agencies to combine experience with expertise in formulating policies that would win public acceptance. In practice, as we have seen, the era of risk regulation coincided with changed public expectations of government, heightening demands for protection from harm along with demands for transparency and reasoned accountability in decisionmaking. The credibility of regulators weakened under these pressures. Charges of inefficiency, bureaucratic rigidity, technical incompetence, and even bad faith proliferated.[46] Judicial responses to the ensuing conflicts seemed to get bogged down in formalism, while the initiative for regulatory reform shifted to Congress and, ironically, even to the scientific community.

Two post-*Chevron* decisions of the D.C. Circuit exemplify the turn toward formalism. *Natural Resources Defense Council v. EPA*[47] challenged EPA for withdrawing a proposal to regulate vinyl chloride emissions as a hazardous air pollutant. EPA had determined in 1976 that vinyl chloride was a human carcinogen for which it was impossible to establish any definite threshold of safety; accordingly, the agency had agreed with environmentalists to press for stricter controls, with an ultimate goal of zero emissions. These concessions were incorporated into a regulation proposed in 1977 but were never issued in final form. In 1985 EPA withdrew the 1977 proposal, explaining that the rule was unreasonably costly and that available control technology would not demonstrably reduce emissions below the then-existing standard. NRDC sued EPA for impermissibly factoring economic and technical considerations into the "health only" mandate of Section 112, the air toxics provision of the Clean Air Act.

The D.C. Circuit could have invoked *Chevron* to uphold EPA's reading of Section 112, but in an *en banc* decision the court read the law so as to award the environmentalists a partial victory. Congress, the court held, had intended EPA to consider health as the primary basis for emission standards under Section 112. EPA's decision to set the vinyl chloride standard on the basis of technological feasibility alone therefore violated congressional intent. The court rejected NRDC's contention that EPA could never consider cost or feasibility under Section 112 but ruled that EPA would have to establish that the contemplated standard was "safe" (that is, adequate to protect the public health with "an ample margin of safety") before taking nonhealth considerations into account. Economic considerations would be appropriate only for selecting among a range of standards that were all "safe" in the sense intended by Congress.

Legislative intent was again the basis for reversal in *Public Citizen v. Young*,[48] a decision involving FDA's attempt to reinterpret the perennially controversial Delaney clause of the Federal Food, Drug and Cosmetic Act. FDA had departed from the Delaney clause's absolute prohibition against carcinogens by permitting the use of two color additives shown to cause cancer in test animals. FDA argued that this policy was justified because both substances posed only *de minimis* health risks. On review, the D.C. Circuit did not question the agency's risk assessment methods. Instead, on the basis of a close reading of the language, context, and history of the Delaney clause, the court held that Congress had firmly resolved to keep even minimally risky carcinogens out of the food supply. There was no room for FDA to carve out an exception for any additive once it had been shown to cause cancer.

In both cases, the court passed up an opportunity to correct political failure, as the Supreme Court arguably had done in the benzene case. Both outcomes, however, were consistent with the adjudicatory philosophy of a conservative era. No longer would there be a court–agency partnership to carry out broad but vague social agendas, working out rule changes and new interpretive conventions such as the "hard look" under the watchful eye of interest groups. Courts would step in only if agency policies appeared to be directly in conflict with congressional intent as judicially construed. If scientific consensus and the law were pulling in contrary directions, as seemed to be the case under the Delaney clause, Congress would have to muster the political will to change the law with no corrective policy guidance from the courts. Whether Congress had the institutional capacity to react and whether the agencies would perform well when granted unsupervised discretion were evidently not to be the concerns of the judiciary.

A different and potentially more threatening formalism made its appearance in a monograph by Stephen Breyer, an eminent federal judge and administrative law expert whom President Clinton subsequently appointed to the Supreme Court. Blaming bureaucratic shortsightedness and public hysteria for the inefficient regulation of small risks, Breyer proposed to create a cadre of specially trained, career civil servants to coordinate risk assessments across the federal agencies and hold them to high standards of technical competence.[49] Like the D.C. Circuit's legislative intent decisions, Breyer's proposal seemed strangely anachronistic. It not only underestimated (as judges so often have done) the contingencies and technical uncertainties of risk assessment, but it greatly overestimated the ability of experts to provide the "right" answers in the increasingly divisive and fragmented politics of risk.

Curiously, something much closer to the vision of Judges Leventhal and Bazelon emerged from the scientific community's growing engagement with risk decisions. Starting in the mid-1970s, industry's complaints of overregulation based on poor science generated a political demand for subjecting administrative uses of science to review by independent expert bodies.[50] Congress enacted several new provisions requiring peer review of the scientific basis for regulation. More important, agencies with well-established review mechanisms—EPA in particular—expanded their reliance on advisory bodies to reduce their vulnerability and improve their credibility.[51] Both Congress and the agencies also called upon the National Research Council, the research arm of the National Academy of Sciences, to assess agency practices and to make recommendations concerning particular areas of risk management.

As if echoing Judge Leventhal's prescription, peer review panels asked questions about the decisionmaking record that were very similar on the surface to those asked earlier by reviewing courts.[52] Was there a reasonable scientific basis for the interpretation adopted by the agency? Had the agency given balanced consideration to the evidence before it? Were its conclusions formulated in a clear, coherent, and comprehensible manner? In sum, in orderly, well-defined regulatory programs, responsible scientific review showed signs of fulfilling much the same function as the "hard look," with the added benefit that the reviewing body had greater technical competence to judge whether the agency had indeed looked hard at the evidence. Appropriately, even the Fifth Circuit, which set a new standard for overly aggressive review in *Gulf South,* later indicated that it would defer to agency decisions incorporating expert peer review.[53]

At the same time, a number of studies by the National Research Council began to reemphasize the value of process in risk-based decisionmaking, much as Judge Bazelon had done a decade or so earlier. A report on risk communication, for instance, stressed the need for interactive exchanges of information and recognized that adequate communication involves multiple messages, both about risk and not strictly about risk, "that express concerns, opinions, or reactions to risk messages or to legal and institutional arrangements for risk management."[54] Reports such as these, authored by the nation's technical elite, pointed to the need for spaces within regulation where experts and nonexperts could interact on quasi-equal terms to address issues on the borderline of science and policy: how to frame the right questions for risk assessment; what information might be relevant; how to ensure credibility; and, not least, when to say that technical inquiry had proceeded far enough to allow a decision to be reached.

Democratizing the Discourse of Government

Though plagued by inconsistency and periodic loss of confidence, the judicial review of science-based regulation left an indelible stamp on the discourse of American regulatory politics. During the expansive 1970s, when both judicial and public interest activism reached new heights, reviewing courts pressed for methods of administrative analysis and explanation that made decisionmaking more transparent and reduced the hegemony of experts. In consequence of aggressive judicial review, the obligation to explain complicated decisions in ways that could be understood by lay judges and the public became an entrenched feature of the U.S. administrative process. Judicial review also underscored both the right and the responsibility of public authorities to decide what counts as valid knowledge for purposes of regulation. In doing so, courts set aside the myth that only scientists could say with certainty when there was a sufficient basis for action.

Whereas cases like *Ethyl* showed a deep judicial grasp of the contingencies of precautionary regulation, the courts were less successful in articulating workable formulas by which judges and administrators could delimit their respective roles as interpreters of regulatory science. The science policy paradigm, in particular, failed to provide usable stopping points for courts taking a hard look at risk assessments. It also permitted the opponents of regulation to question agency expertise through boundary work distinguishing "bad" regulatory science from "good" real science. Judicial withdrawal from the supervision of technical decisions in the 1980s avoided the pitfalls of overzealous review but only by reinstating an unrealistic and anachronistic vision of agency expertise.

In the end, the "hard look" doctrine has proved most durable. Patricia Wald, as Chief Judge of the D.C. Circuit Court, referred to it as a "proud contribution of our circuit" in an address to the American Bar Association's Section of Administrative Law. She observed that the doctrine had subtly changed over time to take account of the court's actual abilities:

> our court now applies it not to tell an agency that its methodology or procedures were wrong . . . as in the days of Judges Leventhal and Bazelon, but rather to tell an agency that it has not sufficiently explained why it chose the course it did . . . Indeed, in just under a third of the direct agency appeal opinions this past year (April 1987–April 1988) in which we reversed or remanded (58 reversals or remands out of a total of 159 opinions), we did so on the basis that the agency's rationale was inadequate. The most common deficiency we find is the agency's failure to explain "departure from prior precedent."[55]

As interpreted by Judge Wald, then, the continued function of the "hard look" doctrine is to force agencies to make themselves understandable to the lay public, even when their actions are based on esoteric and technical information. The obligation is strongest when the agency departs from its own settled practices. That the doctrine was taken up in time by scientific advisory panels is a further tribute to the educational capacity of the courts. The resulting democratization of technical decisionmaking must be seen as a lasting achievement of judicial review, even if the decisions that institutionalized the norm of public accountability were not always wise or beyond scientific reproach.

5

Law in the Republic of Science

Lawsuits involving science and technology typically do not intrude upon the professional activities of researchers or implicate the judiciary in monitoring the internal workings of academic science. Processes that are crucial to governing the "republic of science," such as peer review, funding, teaching, publication, or the day-to-day administration of research projects and laboratories, ordinarily (and scientists would say properly) remain outside the purview of the courts. Yet, like all other matters of substantial public interest or concern, they are not immune to judicial scrutiny. Legal inquiries into research relationships promise special insights into the strategies by which courts at once challenge and sustain the authority of science in civil society, because here perhaps more than in any other setting courts have to confront the discrepancies between science's idealized claims to special status and its actual social practices.[1] Our task here is to determine whether there are areas of scientific inquiry that courts systematically choose not to enter or supervise, and what reasoning supports their decisions not to intervene.

There are two major routes by which courts can become embroiled in the practice, management, or dissemination of scientific research that is not expressly conducted for policy. The first arises through charges of misconduct against scientists and scientific institutions, including claims of fraud, misrepresentation and misappropriation of research results, bias in peer review, and mistreatment of experimental subjects. Such controversies call upon courts to clarify the boundaries of acceptable behavior by scientists and consequently force judicial decisionmakers to probe the normative and social structure of science. A second route is through those rare but often politically salient lawsuits that result from religious or moral opposition to the investigative project of modern science and the privileged status of that project in American society. The attacks on biotechnology by the social critic Jeremy Rifkin, which some have written off as a neo-Luddite repudiation of technology, can be seen as falling in this category (see Chapter 7). Unquestionably moral in inspiration as well are the scattered assaults on science and technology launched in recent years by animal rights activists, conservationists, antiabortion groups, and creationists.

Both types of controversies touch upon fundamental questions bearing on the relation between judicial and scientific authority, and more broadly between science and democracy. When and under what circumstances should judges defer to science's institutional autonomy and capacity for self-regulation? Should the judiciary evaluate challenges to the research enterprise in accordance with standards other than those of the scientific community, or should it defer to science's varied formal and informal codes of self-regulation and professional practice? How should courts balance science's claim to unrestricted freedom of inquiry against claims originating in rival belief systems, such as organized religion? We can approach these issues by briefly reviewing three areas of case law: scientific evaluation and peer review; the treatment of human and animal subjects in research; and the confrontation between science and religion.

Peers in Science, Subordinates in Law

The boundaries of "science" in Western societies have been drawn in such a way that only those inside the boundary—that is, members of the scientific community—are considered capable of judging whether research is meritorious or negligible and whether a researcher's behavior is culpable or deviant. The most important formal mechanism used to maintain this professional autonomy is peer review, science's nearest equivalent to the jury trial; informally, scientists engage in varied forms of boundary work to preserve their disciplinary identity and autonomy.[2] Scientists have long insisted that peer review is the most effective means for allocating grant funds, selecting papers for publication, and determining entitlements to promotion, tenure, prizes, and other professional rewards.[3]

Peer review can take many forms, some far removed indeed from the ideal of neutral and dispassionate scientific inquiry. Scientists in practice freely adopt the rhetorical and dramaturgical practices of the courtroom or theater to expose and discipline problematic colleagues. Thus, mainstream biomedical researchers convened an informal "science court" to discredit a colleague who did not believe that the human immunodeficiency virus (HIV) caused AIDS,[4] and the editor of the prestigious scientific journal *Nature* convened a motley group, consisting of a scientist, a journal editor, and a professional magician, to investigate a French researcher's improbable claims about chemical activity in highly dilute solutions.[5]

Despite demonstrated variations in its motives and procedures, the scientific community often presents peer review to courts and other social institutions as a fail-safe process for evaluating the merits of science (see, for instance, the discussion of *Daubert v. Merrell Dow Pharmaceuticals, Inc.*, in Chapter 3). Yet litigation over science funding, priority setting, disputed authorship, and other types of scientific misconduct has increasingly led courts into examining conflicting images of expert knowledge, professional standards and practices, and personal and professional relations that underlie the impartial facade of disciplinary peer review. Most of these cases present features that would lead us to expect a high level of judicial deference to the forms of self-evaluation used by scientists. Judges are no more knowledgeable about the norms of scientific research—how science should be done—than they are about the substance of scientific evidence. Moreover, as members of an autonomous professional community, judges can be expected to respect another profession's unwritten rules of self-governance. The case law, however, supports a rather more complex reading: courts are prepared to honor science's claims of autonomy, but only so long as they do not conflict with the legal system's major substantive and procedural interests, including the law's own claim to autonomy in finding facts relevant to litigation.

Policing Science: An Ambiguous Record

The treatment of peer review in the courts exemplifies the delicate balancing of scientific and legal authority in recent judicial decisionmaking. Judges on the whole have been much more cautious about second-guessing federal agencies that sponsor scientific research than they are about second-guessing agencies that are responsible for regulating hazardous technologies. Steven Goldberg, a legal scholar, concludes that the "litany of cases provides reasonably precise guidance as to how courts will respond to complaints about science funding decisions: they will respond negatively."[6]

The pattern of judicial deference was set in *Kletschka v. Driver*,[7] the first modern lawsuit against a funding agency. Dr. Harold Kletschka, a heart researcher at the Syracuse Veterans Administration Hospital, asked the Second Circuit Court of Appeals to review a decision by the Veterans Administration (VA) to withdraw a grant it had awarded to him only a year before. Kletschka claimed that the VA's reversal was based on false and slanderous statements made against him by colleagues at Syracuse. The Second Circuit, however, chose to focus on the specialized and technical character of the decision that Kletschka was seeking to reverse. The VA's determination, the court concluded, rested on "subtle and complex evaluations of the technical merits of the plaintiff's project."[8] The court held that a decision of this kind was inherently unreviewable because no judge could hope to match the funding agency's mastery of relevant technical data or its ability to evaluate the researcher's professional competence.

Allegations of discrimination, an area of particular judicial concern, elicited a somewhat more sympathetic hearing in the case of Dr. Julia Apter, although deference to the granting agency's expertise was still decisive. In 1972 Apter, a professor at the Rush-Presbyterian-St. Luke's Medical Center, sued the National Institutes of Health (NIH) on the grounds that her grant application had been improperly rejected because of her gender, her participation in feminist activities, and her testimony against some NIH committee members at a Senate conflict-of-interest hearing. The Court of Appeals for the Seventh Circuit granted that Apter had standing (access to the courts) to pursue this claim, but expressed reservations about the merits of the case. Citing *Kletschka,* the court reaffirmed NIH's broad discretion over research funding.[9] When Apter's case eventually came to trial, she was unable to persuade the court that NIH had unlawfully discriminated against her.[10]

Scientists challenging the priorities set by research funding agencies have fared no better. In *Marinoff v. HEW,*[11] a relatively disinterested plaintiff (not a disappointed grant applicant) sought to compel NIH to support research on a chemical alleged to be a possible cure for cancer. In a brief, dismissive opinion, a federal district court held that it was inappropriate to reverse NIH, since Congress had clearly delegated to that agency decisions about setting priorities in cancer research. Similarly, Kajmer Ujvarosy's suit to force the investigation of a particular theory about AIDS was denied on the ground that the petitioner lacked standing.[12]

But federal judges are not always so respectful of agency priorities where public health is at stake. To take just one, by no means atypical example, in *Environmental Defense Fund v. Ruckelshaus*[13] the plaintiff environmental group challenged the EPA administrator's refusal to suspend the registration of the pesticide DDT. EDF argued that the agency had paid insufficient heed to evidence that DDT caused cancer in laboratory animals and might present a serious risk to human health. The Court of Appeals for the D.C. Circuit responded favorably to EDF's plea, holding that EPA had to provide better reasons for not treating DDT as an "imminent hazard." In this instance, the court seemed perfectly prepared to overrule an expert agency's presumably well-considered regulatory priorities on the basis of contrary scientific assertions by environmentalists. Yet the petitioners in *EDF v. Ruckelshaus* came to court with fundamentally the same complaint as Marinoff and Ujvarosy: that a federal agency with important public health responsibilities had not been sufficiently responsive to specific scientific findings about the control of cancer.

Why should a court defer to NIH's expertise but not to EPA's? There were, to be sure, differences in the social and political context of these decisions. In the DDT lawsuit, a newly organized environmental movement was looking for judicial support in breaking down traditional bureaucratic barriers to regulation. EPA, the target agency, was a newcomer to government and lacked a credible track record in assessing and managing risk. Perhaps more important, Congress had expressed special concern about cancer in the well-known Delaney clause of the Federal Food, Drug, and Cosmetic Act. The D.C. Circuit construed this provision as placing "a heavy burden on any administrative officer to explain the basis of his decision to permit the continued use of a chemical known to produce cancer in experimental animals."[14] The quality of a regulatory agency's explanations, as we saw in Chapter 4, was an issue that courts adopted as their particular province in the 1970s. Nevertheless, by applying different standards of review to issues in research science and regulatory science, courts have helped to create (or perpetuate) divergent expectations about the credibility of experts in academic as opposed to political settings.

Challenges to the confidentiality of peer review and of unreviewed research results have elicited similarly divided responses. Courts have been careful to preserve the confidentiality of deliberations within committees of the National Academy of Sciences and the National Research Council. Thus, the Academy is not regarded as an agency subject to the Freedom of Information Act or the Federal Advisory Committee Act,[15] and in litigation requesting disclosure of Academy drafts and other documents, at least one case denied access to these background materials.[16] Other courts have refused to intrude into the peer review process for fear of disturbing the confidential relationship between government agencies and their expert advisers.[17] Yet when the Equal Employment Opportunity Commission insisted on seeing the peer reviewers' comments in an employment discrimination case at the University of Pennsylvania, the Supreme Court unhesitatingly ruled that the government's interest in preventing race and sex discrimination took precedence over the university's interest in protecting the confidentiality of peer review.[18] In another nondeferential move, the Court in *Daubert* refused to accept peer review as the definitive test for determining the admissibility of expert evidence, stating, "Publication (which is but one element of peer review) is not a *sine qua non* of admissibility; it does not necessarily correlate with reliability."[19] One effect of this ruling may be to make peer review itself more frequently a focus of legal inquiry, thereby undercutting claims by scientists that they alone can determine the legitimacy of this process.[20]

Although courts in the past have taken pains to shield researchers from excessive scrutiny, holding, for example, that a scholar's preliminary results are not subject to disclosure[21] and that the confidentiality of a researcher's relationship with his sources can be protected,[22] more recent cases show a trend toward firmer assertion of the legal system's interest in discovery. Pursuant to requests by litigants, courts have compelled scientific and social researchers to disclose data before publication as well as to provide backup raw data for articles already published in peer reviewed journals.[23] Requests to protect the confidentiality of research subjects may be honored in such cases, and litigants may be asked to compensate researchers for their time, but refusal to surrender the desired information exposes researchers to possible financial and criminal penalties. In opening up the basis for published scientific conclusions, courts implicitly have accepted the possibility that observations are susceptible to multiple interpretations and that the research scholar's own interpretations are not necessarily controlling. Thus, in the guise of maintaining the integrity of legal discovery, courts are willing to tolerate challenges to expert assertions that they have ruled out as inappropriate in other contexts.

Judicial interventions into scientific misconduct cases also affirm the power of courts to superimpose legal norms of acceptable conduct upon those of the scientific community. By the late 1980s scientific misconduct had evolved into a significant public problem. First came a series of embarrassing revelations concerning fraud and plagiarism by promising young scientists at some of the nation's leading research centers. Then, in 1987, the British journal *Nature* published a controversial article written by two NIH scientists who charged that careless misrepresentation, if not outright fraud, was endemic in scientific research and publication. Finally, in 1988–89, two congressional subcommittees chaired by Democratic Congressmen John Dingell of Michigan and the late Ted Weiss of New York held hearings to examine the allegation that a scientist in the laboratory of MIT nobel laureate David Baltimore had misrepresented data in an article published by the prestigious journal *Cell*.[24]

Both the National Science Foundation and the Public Health Service (PHS) promulgated comprehensive rules governing the detection, investigation, and sanctioning of misconduct by their grantees.[25] Initial attempts to apply these rules, however, led to some notable reversals for the government. In *Abbs v. Sullivan*,[26] a scientist's plea for summary judgment against PHS was granted on the ground that the procedures governing the investigation had been improperly promulgated. In a yet more visible and embarrassing reversal, NIH's Office of Research Integrity was forced to drop its multiyear investigation of Robert Gallo, codiscoverer of the AIDS virus, and his colleague Mikulas Popovic after a court determined that the latter had committed no wrongful acts.[27]

These cases may be contrasted with the Second Circuit's disposition of a copyright infringement suit in *Weissman v. Freeman*,[28] in which due process concerns were not at issue. Heidi Weissman, an instructor in radiology at Montefiore Medical Center in New York, had sued her supervisor, Leonard Freeman, for misuse, under his own name, of a "syllabus" or review article on gall bladder imaging techniques of which she claimed to be the sole author. In dismissing Weissman's complaint, the lower court attached much significance to the status difference between the two scientists and implied that Freeman's seniority rendered his authorship claim, if anything, stronger than Weissman's.[29] On appeal, the Second Circuit easily disentangled the legal construct of *authorship* (under copyright law) from the social construct of *authority* (based on the parties' relative scientific status). The appellate court found, on the basis of the evidence, not only that Weissman was the sole author of the syllabus but that Freeman had tacitly accepted her claim.

The Outer Bounds of Research

Courts have been most aggressive in policing the boundaries of acceptable research when the claimant is a human patient with demonstrable injuries. In other circumstances, such as in dealing with the claims of animal rights activists, the judicial response has been markedly more sympathetic toward scientists.

Research with Human Subjects

The sorry record of human experimentation since the nineteenth century reveals many instances in which, in the name of scientific progress or benefits to humanity, researchers overlooked or set aside grave risks to their patient-subjects. In the United States, as in other scientifically advanced nations, the unequal relationship between experimenter and subject led to massive abuses. Some of the most vulnerable segments of society—ethnic minorities, children, women, the sick, the mentally incompetent, the poor and elderly, prisoners and soldiers—were drafted into risky experiments of questionable intellectual or humanitarian value.[30]

The doctrine of informed consent, which today provides the primary line of defense against such exploitation, emerged only after medical science began to differentiate itself from routine therapy. Historians trace informed consent back to an English court decision more than two hundred years ago.[31] Until the middle of the twentieth century, however, courts had relatively little use for the concept, since all deviations from accepted medical practice were viewed with suspicion. "Experimentation" was equated with quackery, and physicians were held strictly liable for harm caused to patients through novel forms of treatment.

By the middle of this century, medical research had come into its own as a respected scientific activity.[32] Concurrently, the egregious crimes committed by Nazi doctors during World War II forced Western societies to consider how best to protect the rights of subjects in biomedical research. The trial and conviction of participants in the concentration camp experiments led to the Nuremberg Code, a monumental statement by a court of law concerning the ethical principles that should govern research on human beings. The Code's first and most widely cited principle is that "the voluntary consent of the human subject is absolutely essential."

It took another decade or two of horrifying revelations to translate the abstract principle of informed consent into legally binding regulations in the United States. Unethical practices were shown to be rampant not only in private research but also in science done by and for the state. For example, in the notorious Tuskegee Syphilis Study, the Public Health Service deliberately withheld treatment from hundreds of black men suffering from syphilis, with the result that at least 28 and perhaps as many as 107 died.[33] The discovery of penicillin in 1929 had provided an effective treatment for syphilis, but although the study was carried out from 1932 to 1972, the participants were never informed that their condition might be treatable.

These disclosures led both federal and state governments to adopt enforceable informed-consent requirements for research with human subjects.[34] As a result, the initial jurisdiction over most informed-consent controversies passed from a judicial to an administrative forum. Government regulation preempted the field to such an extent that only in extraordinary circumstances were controversies involving human experimentation likely to provoke litigation or require a final disposition by the courts. Under regulations issued by the Department of Health and Human Services (formerly the Department of Health, Education and Welfare) regulations, institutions receiving federal research funds must maintain institutional review boards to approve studies involving human subjects and to ensure compliance with consent requirements. Violations may lead to serious sanctions, as they did for Martin Cline, a medical researcher at the University of California at Los Angeles (UCLA).

Cline in 1980 attempted the first documented experiment in gene therapy by inserting recombinant DNA into the bone marrow of two thalassemia patients, one in Israel and one in Italy.[35] The study had not been approved by NIH, his primary sponsor, or by the UCLA Human Subject Protection Committee, which indeed eventually disapproved the study as premature. An NIH disciplinary review forced Cline to resign his division chairmanship, terminated two of his grants, and subjected him to other special penalties.

Informed consent provided a doctrinal basis for post hoc sanctions in the Cline case, but practicing physicians as well as legal analysts remain skeptical of its capacity to change the balance of power between doctor and patient in therapeutic settings.[36] Cases in which courts have drawn the boundary between research and therapy so as to preserve the physician's authority provide additional grounds for skepticism.

Scope and Adequacy of Consent

Apart from threshold questions about whether consent was required or not, conflicts over consent often center on the adequacy of the information given to the patient. A recurrent issue for courts is whose viewpoint to adopt in deciding how much should be disclosed to patients about their medical treatment. The majority of states favor the experts, holding that physicians are responsible for disclosing only as much as would be considered reasonable by a "reasonable medical practitioner" in the same community and the same specialty.[37] This approach is grounded in the so-called therapeutic privilege, which recognizes the physician's preeminent right to withhold any information that might harm the patient. The less deferential minority rule holds that the adequacy of disclosure should be judged from the standpoint of the "reasonable patient," not from that of the "reasonable physician."[38] Although these general rules are well settled, questions about the adequacy of disclosure still arise. For example, in *Arato v. Avedon*[39] the Supreme Court of California had to decide whether a pancreatic cancer victim should have been informed of the statistical life expectancy for patients such as himself and, secondarily, whether the professional or the patient viewpoint should govern this decision.

In the context of experimentation, where standards of medical practice are still fluid and the benefits and risks of treatment are relatively uncertain, the patient-based standard of disclosure seems especially apposite.[40] Abuses of the doctor–patient relationship are less likely to occur if the researcher has to evaluate the adequacy of disclosure from the patient's standpoint. A frequently cited Canadian case reached precisely this conclusion, holding that researchers have a duty to disclose that is "as great as, if not greater than," that of physicians in general.[41]

How a court draws the line between research and therapy may determine, as a practical matter, the degree of deference shown to a physician's judgment in specific cases. *Karp v. Cooley,* an early informed consent case, dealt with this boundary question. Dr. Denton Cooley, a pioneer in heart transplant surgery, attempted in April 1969 to use an artificial heart as a bridge to a heart transplant in Haskell Karp, a chronic cardiac patient. Karp lived with the mechanical device for about three days, but died a day after it was replaced by a donor heart.[42] Karp's widow later sued Cooley, alleging that the validity of Karp's consent to the procedure should have been judged by the higher standards of disclosure applicable to human experimentation.[43] The court concluded, summarily in the view of some,[44] that "the record contains no evidence that Mr. Karp's treatment was other than therapeutic."[45] Karp was judged to have been properly informed by the traditional malpractice test of determining whether a reasonably careful and prudent physician would have acted in the same way under similar circumstances. Twenty years later, George Annas criticized the courts for not using the research-therapy boundary more effectively to control the Defense Department's use of drugs on troops in the Persian Gulf War without their consent.[46]

Whether the situation is deemed therapeutic or experimental, courts are prepared to renegotiate the scope of informed consent so as to offer a remedy to victimized patients. Decisions favoring the patient are most likely when the injury is seen in any sense as preventable. In *Burton v. Brooklyn Doctors*

Hospital,[47] a New York court overrode a physician's judgment and awarded damages to Daniel Burton, an infant plaintiff blinded by excessive exposure to oxygen in a clinical study. At the time of Burton's birth, experts suspected that prolonged exposure to oxygen in premature care might lead to blindness, but this was not "known" as a medical certainty until a year or so later. In fact the pediatric department of New York Hospital, where Burton was treated, had decided about two weeks before his birth that it would participate in a national study to generate definite information about oxygen-blindness. Although researchers saw the issue as unsettled, the jury concluded that the doctors should have "known" not to apply the disputed treatment to Burton; the court agreed, indicating that it was not necessarily "sound medical practice" to follow this treatment for otherwise healthy babies in 1953, "even though it was a common practice at the time."[48] As with the glaucoma patient in *Helling v. Carey* (see Chapter 2), the legal system here wrote its own history of who should have known what, and when, in order to serve its sense of justice.

A propatient outcome is also likely when the complaint is not inadequacy but complete absence of information. Thus, in a study by the University of Chicago and the Eli Lilly drug company, about 1,000 pregnant women were given diethystilbesterol, a drug used to prevent miscarriages, without their knowledge or consent. Although the patients suffered no physical injuries, the drug increased the risk of cancer and reproductive abnormalities among their children. A federal district court permitted the women to sue the university, holding that under Illinois law intentionally administering a drug to an unconsenting patient could be regarded as a battery (an unauthorized touching of the person).[49]

Another case that dramatically expanded the scope of informed consent involved the commercial use of a patient's tissues and cells by medical scientists at UCLA. John Moore, a Seattle businessman, came to UCLA to be treated for a condition known as hairy cell leukemia. His physicians, who were also researchers at the university, discovered that Moore's surgically removed spleen contained unique cells called lymphokines that could be used to manufacture pharmaceutical products of enormous potential value. Without informing their patient, the UCLA researchers established and patented a genetically engineered cell-line from Moore's cells and signed lucrative contracts with two pharmaceutical companies to further develop the cell-line and its products.[50] The market value of products developed from the Moore cell-line was predicted to be approximately $3 billion by 1991.[51]

When Moore discovered these activities, he sued the university and the individual researchers, claiming that the cell-line was his property, that it was unlawful to use it commercially without his consent, and that he was entitled to damages. The California Supreme Court decided that granting patients a property right in their cells would place too great a burden on biomedical researchers, but that Moore nevertheless had a claim based on an expanded theory of informed consent. Specifically, the court held that in a therapeutic situation physicians have a duty to disclose to their patients any commercial interests that might bias or otherwise influence their choice of treatment. Moore could recover for the violation of his autonomy as a patient but not partake in any commercial gains derived from the use of his cells.

The extension of informed consent served here to close off challenges to traditional notions of invention and intellectual property rights while affording patients qualified protection against being unknowingly used as commercial objects. In *Moore,* California's high court signaled in effect that, although medical research was "molecularizing" the patient, the law was not yet prepared to conceptualize the patient's body in a similarly deconstructed fashion for purposes of creating a new property right in genetic information. James Boyle has ably described the court's multiple contortions and contradictions in *Moore,* which present in sum almost a textbook example of how not to write an appellate opinion.[52] Yet the subsequent silence on this issue in other circuits implies that the decision has achieved at least a temporary stabilization of claims having to do with ownership of a research subject's tissues and cells. An unedifying exercise in legal reasoning seems in this instance to have struck a successful ad hoc accommodation between concerns for a patient's rights and widely shared views about the political economy of research, including the idea that society should bestow unconstrained rewards on its most ingenious inventors.

Research with Animals

Some twenty-six times from 1985 through 1987, underground animal rights groups undertook guerrilla actions to free research animals and vandalize laboratories where animal research was in progress.[53] The damage to research projects and facilities was substantial. For instance, in April 1985 the Animal Liberation Front raided biology and psychology laboratories at the University of California at Riverside and carried off 467 animals.[54] Two years later a fire at a veterinary laboratory at the Davis campus of the University of California caused $3.5 to $4 million in damages.[55] The groups that claimed responsibility for such actions—People for the Ethical Treatment of Animals (PETA), Band of Mercy, True Friends—spoke with missionary zeal of "liberating" animals and of "breaking the species barrier." Many were prepared to engage in criminal violence to publicize and promote their goals.[56]

The direct action strategy of animal rights activists, reminiscent of the violent social movements of Europe and Japan, struck a dissonant chord amid the usually orderly hum of protest politics in America. Unlike environmental, labor, and civil rights organizations, animal rights groups rarely turned to the courts, and then mainly as an adjunct to their actions outside the law. In return, they confronted a more restrictive legal environment than most other U.S. social activists, including until recently even the violent wing of the antiabortion movement.[57] Federal courts proved markedly resistant to complaints about the treatment of animals in biomedical laboratories.

Animal rights advocates have sought access to the courts by asserting either that they personally suffered injuries from the mistreatment of research animals (for example, as taxpayers contributing to research) or that they were entitled to serve as legal representatives of creatures that could not otherwise defend themselves. Neither approach has won much favor. The

Ninth Circuit Court of Appeals denied standing to a group that had challenged the treatment of some goats in a federal enclave because the petitioners could claim no direct involvement with the animals.[58] Challenges to an NIH-sponsored research project in Maryland were also eventually dismissed for lack of standing. In 1981 Alex Pacheco, a well-known animal rights activist and founder-director of PETA, volunteered to work for Dr. Edward Taub of the Institute for Behavioral Research. While there, Pacheco compiled evidence that the monkeys used in Taub's research were not being properly cared for. His evidence prompted Maryland prosecutors to bring criminal charges against Taub for failure to comply with the state's animal cruelty law. Taub was tried in the Montgomery County district and circuit courts and was convicted of one count of cruelty to animals. The monkeys were removed to an NIH facility for care and custody.

The Maryland Court of Appeals reversed the lower court judgments on the ground that the state law under which Taub had been convicted did not apply to "a research institute conducting medical and scientific research pursuant to a federal program."[59] PETA representatives attempted in the meantime to gain legal custody of Taub's research monkeys, but the federal district court for Maryland denied them standing. The Court of Appeals for the Fourth Circuit concurred, citing the possible negative impact of litigation on research:

> It might open the use of animals in biomedical research to the hazards and vicissitudes of courtroom litigation. It may draw judges into the supervision and regulation of laboratory research. It might unleash a spate of lawsuits that would impede the advances made by medical science in the alleviation of human suffering. To risk consequences of this magnitude in the absence of clear direction from the Congress would be ill-advised.[60]

The federal Animal Welfare Act, the only guidance given by Congress, appeared in the court's view to subordinate animal protection to "the hope that responsible primate research holds for the treatment and cure of humankind's most terrible afflictions."[61] With courts defining scientific and medical progress as the primary objectives of the law, recourse to the legal process held little hope for the animal rights movement.

Science against Religion

The most direct attack on the content of science has come from religious groups wishing to block the teaching of evolutionary theory in public schools. For American creationists, state legislation was a reliable and malleable ally in efforts to limit the impact of science on children's education and religious values. But the movement's successes in the legislative arena were largely undercut by opposition from the federal appellate courts.

The creationists entered into legal hibernation for nearly forty years following the notorious "monkey trial" of 1925,[62] in which John Scopes was tried and convicted under a Tennessee law that forbade the teaching of evolution. Ronald Numbers, a historian of science and medicine, has interpreted this silence as a sign that the movement had gained success by other means, not that it had disavowed its basic agenda.[63] In any event, by the late 1960s the growing incursion of evolutionary biology into high-school textbooks again energized the creationists into mounting an aggressive new campaign of legislation and litigation.

The Supreme Court in 1968 invalidated a state law similar to the one involved in the Scopes trial on the ground that it advanced the views of a particular religious group.[64] The creationist revival of the 1970s accordingly adopted a different legislative strategy—the so-called balanced treatment approach. The object this time was not to ban the teaching of evolution, but to insist on equal time for both creationism and evolution in the high-school curriculum. Over a decade, several states enacted laws requiring that evolutionary theories of the origin of life on earth not be taught except when accompanied by their creationist counterpart, now labeled "creation science." As in the 1920s, the American Civil Liberties Union took up the challenge posed by these statutes; inexorably, the disputes led to the Supreme Court. In *Edwards v. Aguillard*,[65] the Court ruled by a seven-to-two majority that the Louisiana equal-time law impermissibly advanced religion and hence was unconstitutional.

The high court's decision was perhaps a foregone conclusion, given that the Fifth Circuit had already ruled against the creationists. But the scientific community was taking no chances. Virtually every major scientific organization in the country and at least seventy-two Nobel laureates in the sciences filed *amicus curiae* (friends of the court) briefs urging the Supreme Court to affirm the lower court's judgment that the Louisiana law was indeed unconstitutional. One strategy that the briefs advocated was to demonstrate by sophisticated philosophical argument that "creation science" did not meet any of the uniquely defining characteristics of "science." In a minitreatise on the history and philosophy of science, the National Academy of Sciences sought to prove that science's "naturalistic, testable and tentative" explanations of the world deserved an altogether different kind of deference from the "supernatural, untestable and absolutist" explanations offered by the creationists.[66] Science, in other words, was special in its ability to provide access to the truth.

The Supreme Court gave the nation's leading scientists the outcome they had asked for, but (in contrast with its approach in *Daubert*) the Court chose to stay away from definitions of what constitutes science. Instead, the majority applied a three-pronged test of constitutionality developed in an earlier First Amendment case, *Lemon v. Kurtzman.*[67] Refocused through the lens of *Lemon,* the key question for the Court was whether the label "creation science" was simply a cover for a set of religious beliefs. For answers, the Court turned to the Louisiana statute's legislative history and the backdrop of "historic and contemporaneous antagonisms between the teachings of certain religious denominations and the teaching of evolution." A look at these social realities led the majority to conclude (over Justice Scalia's vocal dissent) that creationism's central tenet—that "a supernatural being created humankind"[68]—was nothing more than a reflection of religious dogma.

Observers have noted that in *Edwards v. Aguillard* the Supreme Court voted more *for* science than *against* religion.[69]

But, equally, the Court reaffirmed its own right to police relations among science, religion, and the state. For *Edwards* was decided without explicitly acknowledging science's claims of cognitive superiority over other forms of belief, as asserted in the *amicus* briefs. Evolutionary theory won not because it was "true" (a position that the dissenters vehemently contested) but because the Court could rely on its own precedents and could find evidence of religious intent in the sloppy legislative tactics of the creationists. Perhaps predictably, *Edwards* failed to quell the *fin-de-siècle* resurgence of struggles between the religious and scientific worldviews in America. Some religious fundamentalists responded to their setback by turning on its head the argument that "creation science" was a form of science. They now claimed that "secular humanism," the complex of values associated with modern science, was really a "religion" in its own right; teaching these values deprived their children of the freedom to practice their own faith.[70] Courts remained generally unmoved by these claims, but away from the legal system fundamentalists won important battles in the campaign to keep evolution out of biology textbooks. Local and regional opposition to science's intellectual and moral hegemony ran too deep in these cases to be thwarted or diverted by the rulemaking machinery of constitutional law.

A Limited Autonomy

Compared with some of the controversies reviewed in earlier chapters, challenges to the internal functions of science have been infrequent, ad hoc, unsystematic, and unpredictable. Unlike governmental experts at EPA and other regulatory agencies, scientists in the laboratory or research setting have few organized adversaries actively seeking opportunities for litigation. When disputes arise within academic science, all those involved in the proceedings—universities, granting agencies, individual scientists, judges themselves—appear committed to resolving them as far as possible without recourse to the courts.[71] Scientists, in other words, actively patrol their boundaries against incursions by the law, citing the efficacy of their self-regulatory mechanisms. Complaints such as Julia

Apter's or Heidi Weissman's seldom reach the courts, and when they do, courts are not especially eager to encourage litigation.

Nevertheless, even this rather unsystematic pattern of interactions sheds some light on the way in which courts view science's claims to professional autonomy and special cognitive authority. In spheres ranging from funding and peer review to research, teaching, and publication, courts have demonstrated that they will defer to the working rules of science only when they do not interfere with the rules or goals of the legal process. Above all, courts are institutionally disinclined to bow to the scientific community if the result will shield practices, documents, claims, or agreements from independent judicial scrutiny in the course of routine legal proceedings.

Courts show little hesitation in piercing the veil of peer review when the objective is to further the adjudicatory process. Deference to the institution of peer review in early cases against funding agencies has given way more recently to aggressive demands for unpublished data and information on research subjects. The Supreme Court's decision in *Daubert* was hailed by some scientists as a welcome invitation to judges to act more like scientists, in part by accepting the greater credibility of peer-reviewed science. But *Daubert,* as seen above, could turn out to be a double-edged sword, for it both denies the absolute reliability of peer review as a marker of "good science" and permits trial courts to reevaluate for themselves the validity of peer review claims in specific cases.

Another area in which the law has declared its normative supremacy over science concerns the research community's duties toward human subjects. Although judicial involvement is not commonplace (since standards of appropriate conduct are determined largely by statutory and administrative rules), courts will insist on full observance of informed-consent requirements when asked to intervene, demanding if anything a higher standard of disclosure from researchers than from treating physicians. Moreover, as the *Moore* case suggests, the perception that a patient is being used for commercial gain may tilt the balance of judicial sympathy against researchers, even if the work itself promises to alleviate human misery.

Yet for all their willingness to probe the hinterlands of scientific practice when their own institutional interests are at stake, courts remain reluctant to participate in frontal assaults on the belief system of science or its methods of acquiring knowledge. Pleas by animal rights activists have produced little in the way of new jurisprudence. Confronted with questions about the moral status of animals, courts have announced that additional social controls should come as needed from legislatures and implementing agencies, not through creative development of the common law. Most tellingly, the higher federal courts have refused to adjudicate between science and religion in ways that could obstruct the spread of scientific knowledge or related secular values. On this contested terrain, science still holds its place as a separate power, whose claims need not be deconstructed with the same skeptical intensity as the motivations of the legislators and interest groups who challenge its cognitive authority. However widely the law may cast its regulatory net over scientific practice, it seems that on the plane of ideology science has little to fear from the courts.

6

Toxic Torts and the Politics of Causation

Toxic tort cases, and the reform proposals they have generated, provide an excellent site for observing how legal norms governing responsibility for technological failures evolve in conjunction with legal ideas of what counts as authoritative scientific knowledge. Courts have rarely drawn as much fire for their supposed misuse of scientific information as in adjudicating toxic tort claims. Cases like that of the infant Katie Wells in *Wells v. Ortho Pharmaceutical Corp.* add compelling specificity to charges of a know-nothing legal system that lets plaintiffs walk away with multimillion-dollar awards although their causal arguments are grossly inadequate by scientific standards. Corporate representatives claim that legal inconsistency and haphazard standards of proof contribute to a climate of uncertainty, motivating companies to make settlements with possibly undeserving claimants. Defendants capitulate, by this account, not because they believe they actually caused harm but because they are unsure what to expect if the case goes before a jury.

Accusations of scientific incompetence intersect in the domain of toxic torts with broader questions about the ability of courts to provide fair, impartial, and cost-effective remedies to victims of technological accidents. Toxic tort litigation has carried forward the general trend in judicial policymaking noted in Chapter 2 toward loosening liability rules in favor of plaintiffs. Conservative economic and political analysts argue that the resulting rule changes—introduced case by case, without extensive consideration of their possible impacts— have created a costly bias against innovation in some sectors of the chemical industry. The remedies most often proposed are either to restrict access to the courts or to bring judicial assessments of science into conformity with "mainstream science."[1]

Plaintiffs' groups, environmentalists, and trial lawyers offer a very different assessment of the tort system's performance, arguing that, despite its flaws and occasional errors, tort law delivers a more effective combination of compensation and deterrence than any other remedial approach—and then only by undercompensating those with the severest injuries. Tort litigation is praised for its flexibility and even for its ability to generate new scientific knowledge relevant to disputes. It is not the technical illiteracy of judges that supporters find troublesome, but the limitations of case-by-case litigation in dealing with the complexities of mass accidents and disasters, which are increasingly commonplace in our technology-intensive society. Consequently, for defenders of the tort system, the real intellectual challenge is to design better procedures for meeting the needs of mass tort victims—for example, by translating the principles of tort liability into a quasi-administrative approach to compensation.[2]

The continuing political controversy about the place of toxic tort litigation in American life underscores the difficulty of evaluating legal policies concerning science and technology in a value-neutral manner. Support for or opposition to the current system inevitably signals commitment to a more complex set of values concerning the rights and responsibilities of individuals and corporations and the optimal way to apportion financial and moral blame for harms inflicted by technology. Any change in the process of litigation has the potential to affect the distribution of winners and losers on a scale that far transcends the immediate controversy. In terms of political meaning, as well, the citizen plaintiff and the corporate defendant stand as symbols and standard-bearers for sharply divergent social and economic philosophies. Accordingly, any legal reform that substantially alters either side's expectations of success is infused with divisive ideological overtones. Values are engaged even when proposed rule changes concern such seemingly apolitical issues as selecting expert witnesses, admitting evidence, or altering the burden of proof.

This chapter does not seek to resolve the debates about the tort system writ large, nor even that part of the system which deals specifically with toxic torts. Michael Saks's compendious survey of empirical research on the tort system strongly suggests that the data that might motivate informed policy changes simply do not exist.[3] Instead this chapter undertakes the more modest but still important task of exploring how authoritative "facts" are constructed within the divisive ambiance of toxic tort litigation. We will identify the processes by which courts assign credibility to competing knowledge claims and participate in erecting boundaries between "mainstream" and peripheral science.

The Birth of a Legal Dilemma

The noted British anthropologist Mary Douglas and the eminent American political scientist Aaron Wildavsky argued in their influential book on risk that disputes over toxic substances exemplify a reversion to primitive patterns of blaming in technologically advanced societies.[4] From the priestly interpreter of the entrails of sacrificial animals to the modern forensic scientist, experts have always been courted for their ability to explain, predict, and avert danger. Modern science and technology, however, have so far improved the mastery of nature that we now seek reasons for calamities that once were treated as mysterious and unavoidable. Previously "natural" conditions such as disease, death, and physical deformity seem strange enough in our materially wealthy societies to require special causal explanations. Lawrence Friedman writes that, along with high expectations of security and comfort, "some significant portion of the population possesses a heightened sense of entitlement, an expectation of possible redress in the face of calamity or injustice."[5] Litigation, already the favored ritual for affixing blame, now provides the means for interrogating science about events that science and technology themselves made unnatural.

Just over two hundred years ago, the English physician Percivall Pott noticed an unusually high incidence of scrotal cancer in chimney sweeps and correctly deduced that the disease was caused by exposure to soot in their work environment. With growing knowledge, other occupational diseases yielded to scientific explanation: for example, black lung disease among coal miners, lung cancer among asbestos workers, byssinosis from exposure to cotton dust, liver cancer in those exposed to vinyl chloride, and sterility among male workers in factories producing the soil fumigant dibromochloropropane (DBCP). Supplementing the observed human responses to toxic agents, a growing body of experimental research established that chemicals could produce diseases and birth defects in animals, especially rodents. Substances known to cause human cancer were found in almost every instance to be carcinogenic in animals.

By the 1970s the new science of ecology began ringing alarm bells about the detrimental effects of toxic chemicals on organisms other than humans and higher mammals. DDT, an effective and widely used chemical pesticide, was banned in most industrial countries upon the discovery that it persisted and accumulated in the environment, left harmful residues in the tissues of crustaceans, mollusks and fish, and interfered with the reproductive cycle of fish-eating birds such as the brown pelican and double-crested cormorant.[6] DDT was also implicated in laboratory studies as a possible human carcinogen and, despite controversy,[7] the suspicion that DDT could cause cancer speeded its banning in the United States. The episode raised public awareness that substances posing risks to one form of life (agricultural pests, for instance) were not likely to be innocuous to other living things.

Disasters involving pharmaceutical drugs also reinforced public awareness that threats to human health and safety could flow from an industrial sector that had previously been regarded as a prime source of "better things for better living."

The thalidomide tragedy, whose impacts were felt more in Europe than in America, nevertheless provided the impetus for a major overhaul of U.S. drug safety legislation in 1962. Other drug-related mishaps, such as those from diethylstilbestrol (DES), swine flu and polio vaccines, Oraflex, and baby aspirin, all confirmed that there are unavoidable risks associated with even the most beneficial pharmaceuticals.

Changes in scientific knowledge made it possible to link chemicals with effects on public health and the environment even in cases in which a causal association could not be definitely established. Techniques for studying the effects of chemicals on laboratory animals gained acceptance as a substitute for assessing risk to humans. Methods for detecting chemicals in the environment and measuring levels of human exposure also developed rapidly, enabling more sensitive determinations of the extent and duration of an individual's contact with toxic substances. Advances in the understanding of molecular and cellular processes expanded the basis for connecting chemical exposure with the induction of diseases such as cancer. Finally, quantitative methods of predicting human risk from indirect evidence—formally known as risk assessment—came increasingly into vogue as a basis for setting environmental standards.

Even in the early 1980s, however, it was not clear just how consequential these developments would one day be for the chemical industry. The Association of Trial Lawyers of America identified "toxic torts" as a distinct area of litigation only in 1977.[8] A 1981 law review article asserting that toxic torts were a "phantom remedy" seemed still in accord with mainstream legal thinking.[9] Yet within a few years toxic tort litigation had moved to center stage in the scholarly and popular literature on the law. Dozens of articles and books, countless press reports, and numerous congressional inquiries focused on a social problem of suddenly intractable proportions. How tort law should deal with scientific uncertainty emerged as a key item for discussion within a wider debate about how to strike an equitable and efficient balance between the producers and the innocent victims of toxic chemicals.

Chemicals and Disease: An Uncertain Connection

A claim that a particular chemical substance or combination has "caused" the plaintiff's injuries lies at the heart of any toxic tort action. Causation here is a special legal construct: it seeks to integrate expert judgments about the plausibility of a particular causal story with the law's normative interest in deciding which kinds of stories are best for individuals and society. Legal rules for deciding under uncertainty, such as the rule that the plaintiff bears the burden of proof, give the clearest evidence of this attempted integration.

To make a persuasive case for compensation, the plaintiff in toxic tort litigation must identify the harmful substance, trace the pathway of exposure, demonstrate that exposure occurred at levels at which harm can result, establish that the identified agent can cause injuries of the kind complained of, and rule out other possible causes. In most toxic tort cases, one or more of these elements is contested, that is, beset by disagreements among relevant communities of experts. The state of knowledge with respect to many toxic agents is extremely imperfect; to satisfy the civil law's "preponderance of the evidence" (or more likely than not) test, what is "known" about a chemical from the general scientific literature almost always has to be supplemented by knowledge acquired about particular individuals and communities of claimants. As we saw in the case of litigation over silicone gel breast implants, knowledge that scientists will accept as both valid and reliable may not be available precisely when it is needed by the courts.[10]

The plaintiff's claim is scientifically strongest when an illness is a "signature disease," uniquely linked to a particular toxic substance. Thus, it is well established that vinyl chloride causes angiosarcoma of the liver, a rare form of cancer; that asbestos causes mesothelioma, a cancer associated exclusively with exposure to mineral fibers of a certain structure; and that prenatal exposure to DES causes a type of vaginal cancer that otherwise occurs extremely infrequently in the general population. In a typical toxic tort case, however, the victim's condition is one that could as easily have been caused by factors other than chemical exposure. It is difficult to establish by a preponderance of the evidence that such commonly occurring complaints as leukemia, birth defects, loss of fertility, and neurological or psychological disorders resulted from contact with one or another toxic substance. Even asbestos victims cannot always demonstrate that conditions other than mesothelioma—specifically, lung cancer and damage to lung tissue—were caused by exposure to asbestos rather than by other causes such as smoking.

The plaintiff's case is usually built by aggregating several types of evidence, none of which would be conclusive on its own. Epidemiological studies are generally the most useful because they focus on human health effects, but methodological defects frequently render such studies unreliable or difficult to interpret. A common failing in epidemiological research is that studies are conducted on small population groups and hence may fail to detect a real, though infrequent, correlation between exposure and illness. For example, if a chemical produces one excess cancer in every 1,000 exposed persons, then a study looking at only 100 individuals would not necessarily discover the correlation. Such studies are said to lack sufficient "statistical power" to establish causal claims about a chemical's effects on health. The technique of "meta-analysis," which allows statisticians to aggregate diverse epidemiological studies to achieve a larger sample size, has generated its own legal controversies.[11]

Epidemiological studies may be inconclusive even if they show a statistically significant correlation between exposure and disease because of a failure to account for possible "confounding factors," that is, factors other than the alleged toxic exposure that could also have produced the observed effects. In a study purporting to link asbestos and cancer, for example, the researcher must properly control for smoking, which can lead to similar kinds of lung disease. More generally, control groups must be selected so as to rule out factors such as diet and age that are often associated with health effects similar to those caused by toxic substances. Care must be taken to ensure that the study subjects were exposed to the suspected chemical at levels and time periods consistent with the known etiology of the disease.

If epidemiological studies often fall short of providing the level of proof required, courts may find other categories of evidence to be even less satisfactory. Chemical structure analyses, *in vitro* cellular studies, and animal studies, for example, are given limited weight in toxic tort cases except in conjunction with epidemiological data. Animal studies are subject to many of the same design flaws as studies of human populations; factors such as inadequate numbers of animals, improper monitoring of dose or diet, and defective laboratory practices can render experimental evidence virtually worthless. By contrast, well-designed and well-conducted animal studies can elucidate the biochemical processes by which a chemical induces disease, thereby providing important support for effects observed in flawed epidemiological studies. Animal data can also provide limited quantitative evidence of risk. Accompanied by only weak epidemiological data, however, such indirect evidence has failed to persuade judges in a range of cases, including the Agent Orange and Bendectin trials.

Even reliable evidence from animal and human studies only partly satisfies the plaintiff's burden of proving causation. Scientific studies can only at best establish to the research community's satisfaction just the issue of "general causation": that the plaintiff's health condition could in principle have been caused by the alleged exposure. The plaintiff must still provide proof of "specific causation": that the chemical did in fact cause his or her specific injuries. This requirement, too, poses a formidable obstacle to recovery. Positive epidemiological studies can be used to show a statistically increased probability that the plaintiff might have contracted a disease as a result of a toxic exposure. But such a statistical demonstration will not by itself meet the preponderance test except in those relatively rare cases in which illness occurs in more than 50 percent of the exposed population. Most toxic tort plaintiffs must therefore provide additional corroborative evidence that their illness actually resulted from the claimed exposure.

A major hurdle in proving specific causation is to show that exposure occurred in amounts and durations sufficient to trigger illness. Exposure is notoriously difficult to characterize with precision, especially in cases involving environmental toxicants. Farm workers, for example, are rarely in a position to establish their exact levels of exposure to pesticides or the exact combination of substances to which they were exposed. Even factory workers, whose environments are generally more carefully monitored for exposure to toxic substances, do not always have access to records establishing the extent and timing of their exposure. Thus, in *Christopherson v. Allied Signal Corp.*[12] a federal court dismissed a chemical worker's suit when the only evidence he could provide in this regard was a co-worker's statement that he had been exposed to unspecified toxic fumes. In reaching this result, the court tacitly held that "objective" data—the chemical composition of the fumes, numerical evidence of plant size and exposure—should take precedence over the co-worker's subjective and experiential knowledge of conditions in the workplace.

Plaintiffs encounter still greater difficulties when exposure occurs through environmental contamination from a hazardous facility or waste disposal site. In such cases, mathematical modeling is often the only available technique for investigating whether a pollutant plume might have contaminated the air or migrated into the water supply. The uncertainties involved in such modeling render its use in litigation extremely problematic, especially in view of the judicial resistance to indirect and statistical forms of proof.

Scientific indeterminacies are compounded in mass tort cases by indeterminacies that flow from the circumstances of production in the modern chemical industry, and indeed from the social organization of knowledge. Frequently there is more than one manufacturer or discharger who might have been responsible for the harmful exposure, and plaintiffs are not in a position to uniquely identify the defendant who injured them. Relaxation of the standard of proof in *Sindell v. Abbott Laboratories* (see Chapter 2) represented a creative if not widely emulated judicial effort to overcome the barrier presented to plaintiffs by indeterminate defendants. The *Christopherson* court, by contrast, was less concerned with the irreducible uncertainties that surround toxic tort claims; in turning aside the plaintiff's offer of evidence, the court did not inquire, for example, whether exposure data of a more "objective" or "scientific" kind were in fact available to Christopherson. Mass torts can produce troublesome indeterminacies for defendants as well, in the form of uncertainty about how many plaintiffs might eventually be drawn into litigation. The issue for courts in these situations is less how to find facts than how to decide who should bear the costs of society's inability to ascertain the relevant facts with any degree of certainty. Fact-finding, therefore, is a normative and deeply political exercise, whether or not it is openly recognized as such by the parties or their legal representatives.

Empiricism in Judicial Fact-Finding

Facts, nonetheless, must formally be found in any toxic tort action, and courts as much as scientific laboratories are places where demonstration remains a powerful form of argument. Demonstrations that carry legal weight, however, are constrained by all the factors discussed in Chapter 3, as well as by interpretive conventions that have evolved in the context of personal injury litigation. When choosing among competing theories of causation, for instance, judges and juries show a marked preference for eyewitness accounts that bring the causal story to life within the courtroom. Medical and clinical data are seen as more persuasive than animal studies or statistical evidence, and parties who wish to rely on quantitative data secure an advantage when their experts are able to make the numbers real and palpable to the fact-finder. In sum, the construction of scientific facts in litigation necessarily proceeds through quite different avenues from the production of facts by scientists for the consumption of their peers.

The Treating Physician Syndrome

Tort law generally defers to treating physicians who have medically examined the plaintiff whether or not they are knowledgeable about all the factors relevant to causation.[13] In

Ferebee v. Chevron Chemical Co.,[14] for example, a federal court of appeals was asked to review a jury verdict for a plaintiff who had died from a lung disease allegedly caused by exposure to the herbicide paraquat. The testimony offered by Chevron, the defendant chemical company, dwelt largely on the issue of general causation: was there any evidence that paraquat could have caused the plaintiff's disease and death? Company witnesses, including a pulmonary specialist, a pathologist, and a radiologist, all testified negatively on this issue.

The plaintiff's legal strategy, by contrast, focused on the issue of specific causation, and hence by extension on making Ferebee himself the center of attention. The jury was shown a videotape of Ferebee describing his medical history and the circumstances of his contact with paraquat. Two physicians who had treated Ferebee offered live testimony, explaining his medical history, his laboratory tests, consultations with other doctors, and the medical literature on paraquat-induced illness. The appellate court concluded that this evidence was sufficient to warrant the jury verdict for Ferebee: "There were expert opinions on both sides of the issue. In such a situation, the weight to be given to each expert opinion is exclusively for the jury. Drs. Crystal and Yusuf were both experts eminently qualified in pulmonary medicine. They were both among Mr. Ferebee's treating physicians, *a qualification shared by none of the defendant's experts*" (emphasis added).[15] The court, in other words, reaffirmed the jury's right to place treating physicians, who bore direct witness to Ferebee's condition, at the top of the hierarchy of credibility.[16]

Medical evidence that is *not* provided by the plaintiff's physician encounters persistently higher entry barriers. In *Boltuch v. Terminix International, Inc.,*[17] a New Jersey court refused to let Dr. Samuel Epstein, a licensed physician, testify about the cause of the plaintiffs' health complaints because he had neither personally examined them nor relied on their medical records. Medical experts, the court concluded, could not "reasonably make such complicated diagnoses of specific individuals' problems without examining or speaking to such individuals."[18]

Similarly, in *Thompson v. Southern Pacific Transportation*

Co.,[19] the court ruled against the plaintiff partly because five physicians, all of whom had treated Thompson, testified that they did not believe his disease had been caused by dioxin. In the Agent Orange opt-out decision, Judge Weinstein indicated that he, too, would have preferred medical evidence obtained from direct examination of the plaintiffs.[20] By contrast, in *Wells v. Ortho Pharmaceutical Corp.,*[21] testimony by the plaintiffs' treating physicians won out over that of the defendant's experts, who included some paid (hence presumably not disinterested) consultants. All these cases suggest that courts are prepared to overlook weaknesses in the showing of general causation when presented with strong proofs of specific causation by medical practitioners. By the same token, convincing arguments about general causation are rejected when the plaintiff fails to make a persuasive showing of specific causation.

Implicit in all these decisions is a privileging of some forms of knowledge over others: to paraphrase Jethro Lieberman,[22] courts seemed inclined in each case to favor a holistic (or medical) to a reductionist, toxicological model of illness. The holistic view focuses on the suffering individual and asks whether, given the totality of circumstances, this person could have been affected in the stated way by the stated exposure. As reflected in the treating physician syndrome, this approach presumes that issues of general and specific causation must be addressed together, within the context of the plaintiff's lived life. This view is sharply at odds with the opinion of some toxic tort critics that general causation must be established *prior to* specific causation. David Kaye, a law professor, expresses the typical bias when he writes, "Generic causation is a necessary but not a sufficient condition for individual-causation."[23] Needless to say, such a formulation unreflectively adopts a judgment by some part of the scientific community about what is an adequate showing of "generic causation." Given the facts in *Christopherson,* for example, how would a third-party observer such as a judicial fact-finder (or, for that matter, Kaye himself) "know" whether "exposure to nickel and cadmium causes small-cellular carcinoma of the colon"? To accept a particular theory of general causation as true, the court must accept, with or without attempted deconstruction, the expert consensus on which that theory is based—in the form in which that consensus is represented in court. Decisionmakers who privilege the knowledge of treating physicians, for their part, seem no more reflexively aware of their hidden normative preferences, although these preferences are biased toward the contingencies of specific or individual causation—that is, toward recognizing that something could be true for an *individual* even if unlikely in general. The approach favored by scientific and legal positivists, by contrast, is to accord more weight to what is statistically probable, as reflected in stories of general causation.

Epidemiology and Class Actions

In cases involving large classes of claimants, epidemiological evidence often influences outcomes in much the same way as testimony by treating physicians in cases with fewer plaintiffs. In *Allen v. United States*,[24] a lawsuit against the federal government by more than a thousand residents of Utah and Nevada for cancer allegedly caused by fallout from nuclear testing, trial judge Bruce Jenkins relied on epidemiological evidence of radiation-induced cancer effectively to shift the burden of proof to the defendant. Once the *Allen* plaintiffs showed that they had been exposed to radiation during the period of atomic testing and that the available epidemiological evidence linked their particular form of cancer with radiation, the judge concluded that radiation had been a "substantial factor"[25] in increasing the plaintiff's risk of cancer. The claimant was awarded damages unless the government could prove that the particular instance of cancer was *not* caused by fallout.[26] An appellate court eventually dismissed the case on the ground that the federal government could not be sued when carrying out discretionary functions such as weapons testing.[27]

More diagnostic of judicial and scholarly leanings was the Agent Orange lawsuit, one of the most controversial mass tort cases of the 1970s, possibly of the century. The case pitted thousands of Vietnam veterans against the manufacturers of a herbicide that had allegedly caused a wide spectrum of health injuries, ranging from cancer and birth defects to severe neurological and psychological abnormalities. Like the great majority of civil suits, the case never went before a jury, but was settled on the basis of lengthy pretrial negotiations supervised by federal district judge Jack Weinstein. Before this point was reached, Weinstein made several important rulings that reflected a special respect for clinical evidence.

The judge's preferences came to light most clearly in his handling of the claims made by about 350 plaintiffs who had opted not to join the larger class of veterans covered by the settlement. When these "opt-outs" resumed litigation, the defendant chemical companies moved for summary judgment, a procedure by which a court agrees to dismiss a lawsuit in which the originating party has no chance of prevailing. Since a summary judgment extinguishes the legal rights of one party before the case is tried, such motions are granted only when there is no disagreement over any facts relevant to the case, and no interpretation of the facts would permit the plaintiff to win as a matter of law.

The opt-out plaintiffs presented evidence from both animal and human studies to support the claim that their injuries had more probably than not been caused by Agent Orange. Judge Weinstein, however, was unimpressed. Appraising the evidence for himself,[28] Weinstein concluded, first, that the epidemiological studies relied on by the plaintiffs were not strong enough to support their claims. Second, he rejected the animal data supplied by the plaintiffs, stating that "inapposite extrapolations from animal studies"[29] were insufficient to establish causal connections of the kind alleged by the opt-outs. These two rulings effectively demolished the opt-outs' scientific arguments and provided the basis for granting their opponents' summary judgment motion. More important, they presaged the marginalization of animal data in toxic tort suits. The position that "mainstream" scientists would like courts to adopt is reflected in the following comments from an article in *Science*: "High-dose animal studies have questionable relevance to risks to humans from low-dose exposures. Such evidence, presented outside the context of a comprehensive risk assessment, is a gross misuse of scientific data that should be excluded from the courtroom."[30] Interestingly, the authors here accept a "comprehensive risk assessment" as capable of validating otherwise unreliable animal studies. This is a curious inversion of the position taken by toxicologists in regulatory controversies, where risk assessment more often is seen as distorting the valid qualitative messages that could be derived from animal studies.[31]

Increased Risks

Suits claiming damages for increased risk also illustrate the judicial system's preference for observable, empirical proofs. In the past, the reluctance of judges to award damages for merely speculative injuries ruled out any possibility of success in such suits. Now, however, accumulating knowledge of the molecular and cellular changes that harbinger disease, together with the proliferation of tests for detecting such signals,[32] has made it possible for plaintiffs to provide physiological evidence of risk long before there is (or could be) a detectable clinical illness.

Scientifically tenuous or not, testimony concerning increased risks can be made to appear tangible rather than speculative in the hands of trained experts. Clifford Zatz, an experienced defense lawyer, expressed misgivings about charts that display immunological test results to juries through "row after row, column after column" of colored dots representing alleged abnormalities. Such "painting by numbers," Zatz contended, can have a tremendous impact:

> This is powerful testimony. It is simple, it is dramatic, and it is unforgettable. It makes allegedly subclinical injury almost visible to a jury that has come to expect a look at the amputated leg, a glimpse of the burnt flesh, a living reminder of a mistake in plastic surgery, or the proverbial x-ray of the surgical tool left inside the body. Practically speaking, it leaves the defense with an awful lot of explaining to do.[33]

The persuasive power of such testimony in the courtroom, then, is to some extent independent of its conformity with widely accepted standards of technical accuracy. Represented by dots in a chart, test outcomes become mute "eyewitnesses" to actual events. Statistical questions, such as whether these results represent significant or insignificant deviations from the "normal" range, and questions about the validity, reproducibility, and clinical significance of tests may fail to reach the jury unless the opposing party is able to raise them in equally compelling form.

Plaintiffs' lawyers see increased risk cases as permitting earlier remedial action (immunological damage, the primary basis for increased-risk actions, occurs much sooner following chemical exposure than long latency diseases like cancer), but chemical manufacturers fear that claims built on persuasive but imperfect data could open up a new front of ruinous liability. As yet, however, these concerns seem unwarranted. Courts have granted some modest collateral claims presented by increased-risk plaintiffs, but they have usually resisted pleas to award present damages without a showing of present harm. In *Ayers v. Township of Jackson*,[34] for example, residents of a New Jersey town brought an action for a variety of injuries—including heightened risk of cancer—caused by toxic pollutants leaching from the town landfill into their water-supply wells. The New Jersey Supreme Court denied damages for the increased risk per se but allowed the plaintiffs to recover the cost of medical surveillance. The case for medical monitoring, the court held, could be satisfied by establishing a significant risk of serious disease. It was not necessary for the plaintiffs to show a quantified, reasonable probability of harm.[35]

Sterling v. Velsicol Chemical Corp.,[36] another increased-risk case, struck a somewhat different balance between defendants' rights and plaintiffs' interests. Evidence at the trial indicated a 25–30 percent greater likelihood of disease for plaintiffs who had been exposed to the defendant's toxic discharges. The Sixth Circuit Court of Appeals held that this was not sufficient to establish "with reasonable medical certainty" that the plaintiffs would contract kidney or liver disease.[37] As in *Ayers*, the court denied recovery for the increased risk itself but concluded that the plaintiffs' fear that they might become ill in the future was compensable under Tennessee law.

A Place for Fear

Increased-risk cases are important not only as a showcase for the evidentiary preferences of judges but also as an example of the central role that courts play in certifying some forms of social resistance and unrest as legitimate enough to warrant a public response. No matter how narrowly they have fashioned the remedy for increased risk, U.S. courts have gone on record to say that it is not unreasonable for people to fear some of the products of industrialization, such as chemicals and power lines. Courts indeed seem consciously or unconsciously to share in these fears. Thus, the Third Circuit Court of Appeals commented in a case involving exposure to polychlorinated biphenyls (PCBs) that "in a toxic age, significant harm can be done to an individual by a tortfeasor"; medical monitoring claims, the court held, were appropriate "in an effort to accommodate a society with an increasing awareness of the danger and potential injury caused by widespread use of toxic substances."[38]

Perhaps less edifyingly, the Court of Appeals of New York ruled in *Criscuola v. Power Authority of the State of New York*[39] that claimants seeking to recover for a loss in the market value of their property around an easement for a power line did not have to prove that the fear of cancer which caused the loss was scientifically reasonable. As noted in Chapter 2, the court in this case was prepared to hear expert testimony from economists on the question of market value but feared that the testimony of scientific experts would impossibly cloud the issue.

While technocratic critics may deplore as groundless such validation of the public's "irrational cancerphobia," a more sociologically sensitive analysis would see these cases as confronting in a potentially productive way the pervasive loss of trust and social anomie that many see as a condition of life in technologically advanced societies.[40] For example, prompted in part by warning signals from the courts, some nineteen states had undertaken as of late 1993 to address the risks posed by electromagnetic fields. Their responses were diverse, falling into four categories: prudent avoidance, moratoria on new transmission line construction, adoption of numerical field strength standards, and preservation of the status quo.[41] Such decentralized and experimental efforts to address cognitive and social uncertainty may in the end do more to create a positive environment for technology than would denial and an uncompromising refusal to engage in public debate. We may recall that consistent support for the proponents of nuclear power from the highest court in the land—including, in *Metropolitan Edison Co. v. People Against Nuclear Energy (PANE)*,[42] a refusal to treat public fear about the restart of the Three Mile Island nuclear power plant as an "environmental impact"—led only to an effective moratorium on what nuclear engineers regard as a safe and efficient source of energy. Judicial endorsement of expert views may close off an important forum for debate on technological issues without significantly enhancing public trust in the reliability of experts and their technological predictions.

Searching for Mainstream Science: The Case of Clinical Ecology

To correct the perceived deficiencies of scientific assessment in toxic tort cases, courts have been enjoined to take their cues from "mainstream science."[43] In Chapter 3 and again in this chapter, I have argued that such "mainstream" scientific positions (if they exist at all) are constructed in part through the very flow of litigation. Knowledge that the tort system may eventually be willing to live by emerges from the dialogic interaction of law and science, often in conflicts about the safety of technology. The case of "clinical ecology," frequently cited as an example of "junk science" by the advocates of "mainstream science," shows this process at work. It suggests that the adjudicatory process is not structurally incapable of taking into account differences between so-called mainstream and peripheral views in science, but that (as within science itself) it takes time and active work by professional bodies to develop consensual knowledge around new areas of dispute.

The gradual elimination of "clinical ecology" experts from American courtrooms suggests that mainstream science can reassert itself in toxic tort litigation through the knowledge-policing efforts of scientists themselves. Clinical ecology is a name used by a loosely organized community of immunologists and allergists to describe a set of beliefs concerning the way chemicals in the environment affect human health. Members of this community ascribe a wide variety of physical and psychic disorders (for example, depression, chronic fatigue, respiratory and gastrointestinal symptoms, and hypertension) to chemical exposure, alleging that environmental chemicals produce "immune system dysregulation" in especially susceptible patients. Clinical ecologists have developed treatments for an illness that "has been variously dubbed multiple chemical sensitivities (M.C.S.), environmental illness, total allergy syndrome or, more dramatically, 20th-century illness."[44] These include drastic modifications in the patient's living and work environments, such as restricted diets and isolation in "safe" rooms with special air filtration systems.

Starting in the early 1980s, clinical ecologists began appearing as witnesses for plaintiffs in lawsuits against chemical producers and dischargers. Their early appearances led to some notable victories for plaintiffs. For example, in *Anderson v. W. R. Grace*,[45] testimony by Dr. Alan Levin, a leading practitioner of clinical ecology, persuaded a federal district court in Massachusetts to deny the defendant's motion for a summary judgment. Levin testified that the cell damage he had observed in the plaintiffs was a precursor to leukemia and could have been caused by Grace's discharges. Grace eventually reached a multimillion dollar settlement with the plaintiffs.[46]

In *Menendez v. Continental Insurance Co.*,[47] the plaintiff's principal witness, Dr. Alfred Johnson, who was both her treating physician and a practicing clinical ecologist, testified that her condition had been "triggered" by an adverse drug reaction that "'sensitized' plaintiff to many things in her environment, including formaldehyde found in carpet, cigarette smoke, and such foods as carrots, sweet potatoes, chicken, walnuts, rice, beef, eggs, certain fruits, kidney beans, soy, red snapper, salmon, shrimp and flounder." The defendant's expert countered that "he did not think clinical ecology was a scientific practice,"[48] but this appeal to science did not prevent the Louisiana court from deferring to Dr. Johnson's medical judgment.

In several later cases, however, testimony by clinical ecologists received much less hospitable treatment from the courts. Thus, in *Rea v. Aetna Life Insurance Co.*, a U.S. district court in Texas dismissed an antitrust suit by a group of clinical ecologists, finding *inter alia* that "clinical ecology is not a board certified specialty of medicine requiring formal training or testing nor identifiable with any set of independently established standards of practice or body of knowledge."[49] Similarly, in *Laborde v. Velsicol Chemical Corp.*, a Louisiana court held that "'Clinical Ecology' is not a recognized field of medicine."[50] Finally, in *Sterling v. Velsicol Chemical Corp.*[51] the Sixth Circuit Court of Appeals held that the opinions of the plaintiffs' experts on immune system dysregulation, including testimony by Dr. Levin, lacked sufficient scientific foundation to prove that water contaminated by the defendants had injured the plaintiffs.

This reversal in judicial receptivity toward clinical ecology must be attributed in large part to a successful marginalization of the self-proclaimed specialty by the orthodox medical community. During the 1980s several influential professional organizations criticized the scientific basis for the claims of the clinical ecologists. In 1985 the American Academy of Allergy and Immunology (AAAI) circulated a position statement stating that "The theoretical basis for ecologic illness in the present context has not been established as factual, nor is there satisfactory evidence to support the actual existence of 'immune system dysregulation' or maladaptation." AAAI therefore labeled clinical ecology "an unproven and experimental methodology."[52]

Another professional organization, the California Medical Association (CMA), concluded as early as 1981 that clinical ecology did not constitute a valid medical discipline. In 1984 CMA convened a scientific task force to reevaluate the question in detail. Chosen "for their expertise in internal medicine, toxicology, epidemiology, occupational medicine, allergy, immunology, pathology, neurology and psychiatry,"[53] the CMA task force could claim with some credibility to be speaking for medicine as a whole. Additional authority derived from the group's procedural choices, which drew upon both scientific and legal strategies of legitimation. On the scientific side, the task force conducted a literature review and adopted explicit methodological criteria; at the same time, in a quasi-judicial vein, it held a public hearing in April 1985, at which proponents and opponents of clinical ecology presented oral testimony to task force members. On the basis of these investigations, the task force concluded that there was no convincing evidence "that patients treated by clinical ecologists have unique, recognizable syndromes, that the diagnostic tests employed are efficacious and reliable or that the treatments used are effective."[54]

These efforts to redraw the boundaries between legitimate and illegitimate medical science soon influenced judicial thinking. Clinical ecology suffered an important setback in the *Velsicol* case, in which the court noted that "the leading professional societies in the specialty of allergy and immunology . . . have rejected clinical ecology as an unproven methodology lacking any scientific base in either fact or theory."[55] Notably, however, the court made no attempt to look behind the consensus generated by AAAI and CMA to question whether these organizations possessed any hidden biases or whether their consensus-building efforts afforded sufficient protection against bias. As far as the court was concerned, the professional societies could simply be accepted as "black boxes" capable of distinguishing sound medical theories from unsound ones. This decision may be applauded from the standpoint of efficiency, but it again illustrates a propensity on the part of appellate courts to adopt and refract back into society the scientific community's perceptions of its own cognitive and social credibility. The law in this way plays a powerful role in upholding the legitimacy of science in society.

Toward Meaningful Policy Reform

The question of how to improve the adjudication of scientific issues in toxic tort litigation cannot in the end be separated from questions about the adequacy of litigation itself as a method of compensating the victims of exposure to toxic substances. Opinions concerning the latter issue, however, remain deeply divided, ruling out the possibility of an easy national consensus on reform. Positions have hardened around two essentially irreconcilable attitudes about the effectiveness of the tort system in its present form. Any movement beyond this point is likely to require deep-seated changes in the nation's legal and political consciousness, including a resolution of the national debate on how to bring about equitable and affordable policies for health care. As suggested by the fate of President Clinton's health care plan in the 103rd Congress, the chances that this will happen through top-down legislative reform are vanishingly small in the political climate of the mid-1990s.

At one end of the tort reform debate are the "incrementalists": those members of the research community, the plaintiffs' bar, and the general public who endorse the adjudication of technical claims in a lay forum and who generally approve of the way the tort system has functioned to compensate toxic substance victims. The incrementalists are satisfied that the tort approach produces more equitable results in the aggregate than do other remedial systems, even if some undeserving plaintiffs receive windfalls and many deserving plaintiffs remain undercompensated. From this perspective, the resolution of scientific controversies in an adjudicatory setting is almost a nonproblem. The incrementalists see the deck as so heavily stacked against most plaintiffs that none but the most deserving cases stand a great chance of success. Saks's study of the tort system provides important support for these opinions. He concludes that

> A tiny fraction of accidental deaths and injuries become claims for compensation; even known actionable injuries rarely become lawsuits. In both federal and state courts, torts has not been the largest or fastest growing area of civil litigation. The great majority of all kinds of civil suits result in negotiated settlements. On average, these settlements undercompensate the plaintiff's losses. Modest losses are fully or overcompensated, but the larger the loss suffered, the more pronounced the undercompensation.[56]

Supporters of the tort process may concede that there is room for refinement in the handling of science, but they are satisfied with a gradualist approach, in which judges, juries, and lawyers develop greater technical competence through mutual education and case-by-case exposure to toxic tort actions.

The situation looks very different to the "radical reformists," primarily scientists, industry representatives, and some academics, who advocate significant institutional changes in the handling of the scientific aspects of toxic tort cases. These proponents of greater professionalism in the tort process emphasize what they view as "unscientific" features of judicial thinking, such as excessive reliance on treating physicians, undue skepticism toward statistical evidence, and disregard of issues relevant to general causation. The professionals are concerned, more generally, about an inadvertent relaxation of standards of proof, not only through mistaken applications of existing rules, but also through the failure of fact-finders to ask essential questions about exposure and intervening causes. Focusing on alleged mistakes in the tort system's assessment of science, the reformists argue that the only way to ensure more scientific responsibility in the courtroom is to delegate a greater share of decisionmaking to experts, for example, through expanded use of special masters, blue-ribbon juries, expert advisory panels, consensus reports and manuals, and better technical training for judges. The overall objective of these reform proposals is to make the theater of the toxic tort action more receptive to that often elusive actor: "mainstream science."

The historical record suggests that the incrementalists have the stronger political case. There has been no dearth of opportunities for national and state legislatures to consider alternative approaches to victim compensation. In the wake of the polybrominated biphenyl disaster in Michigan and the pollution incident at Love Canal, several personal injury compensation bills were introduced before the 96th Congress.[57] The 97th Congress commissioned a detailed study of existing legal remedies, and the resulting report, published in mid-1982, included an ambitious proposal for legislative reform.[58] During the 98th Congress the House Committee on Public Works, through its Subcommittee on Investigations and Oversight, held extensive hearings on the issue of victim compensation.[59] Ironically, this lengthy investigation persuaded the committee that "there is simply *insufficient information* to enable the Congress to properly construct a viable *long term legislative remedy* to the victim compensation issue."[60] From his survey of the tort system as a whole, Saks adds a contemporary note of concurrence: "Even if we agreed on the values and goals of such a system, we have no sound basis for concluding that those goals would be more likely to be reached by abandoning the tort system in favor of a no-fault system, by finding ways to make a tort system that is more active and efficient, or by retaining the current system."[61]

This chapter's central argument—that legally compelling knowledge about the toxic properties of chemicals arises not from science alone but through complex interactions between adjudication and scientific activity—also militates in favor of incrementalism. "Mainstream science" does not exist in a pure cognitive domain that courts can reach into at will. Like all other human knowledge, "mainstream science" is *made*, and it is made in part through the incremental efforts of the legal system to acquire relevant knowledge.

To admit this is not to deny that there are defects in the traditional approach to litigating scientific uncertainty: the sometimes gross mismatch between scientific and judicial appraisals of credibility, the often ill-motivated efforts to fit scientific claims to legal concepts of causation, and the discrepant and idiosyncratic results reached in science-intensive litigation in different jurisdictions. The decentralized character of litigation ensures that improvements in judicial case management take a long time to trickle through the system, and there is painful evidence from the example of asbestos litigation that, in spite of much creative procedural innovation in complex litigation, judicial learning alone is not enough to handle major toxic disasters.[62] More could very possibly be done to speed up the legal system's processes of learning and its capacity to accommodate changing knowledge. We will return to these possibilities in the final chapter.

7

Legal Encounters with Genetic Engineering

The legal developments discussed in this and the next two chapters illustrate the law's central role in constructing public understandings of technology—particularly the risks of technology—more clearly than in the case of toxic torts. We saw in the preceding chapter that liability rules not only altered notions of responsibility and victimhood but also, by recognizing claims based on increased risk, cancerphobia, and clinical ecology, validated to some degree new public attitudes toward chemicals. For the heavily regulated, structurally mature chemical industry, however, there are limits to the new social meanings that can possibly be created or ratified through the legal system. By contrast, the courts were involved in controlling biotechnology when its institutional setting and regulatory status were still very much in flux. One might reasonably have expected their impact at this stage to have been more profound.

Public hysteria, cynical lawsuits, distrust of experts, judicial ignorance and overreaching—all the primal fears that litigation calls forth from the scientific community's collective unconscious did indeed come to the fore as federal courts were drawn into regulating biotechnology. The courts were in the spotlight as soon as they provided a forum to Jeremy Rifkin, the nation's best-known and most effective individual opponent of biotechnology, a man who is not afraid to be called "the most hated man in science."[1] To some a Naderesque hero and to others a Luddite villain, Rifkin showed boundless talent in polarizing public opinion. In April 1986, for example, the *New York Times* applauded his organization, the Foundation on Economic Trends (FET), for exposing serious inadequacies in the federal system for regulating biotechnology.[2] Just a few days later, the *Los Angeles Times* accused Rifkin of cleverly manipulating the courts and the regulatory process "to halt one of the most significant scientific advances of our time." Dismissing his predictions of disaster as groundless, the paper warned that "the only danger to humanity lies in continuing to listen to Rifkin."[3]

Scientists at the forefront of biotechnological research greeted Rifkin's legal activism with dismay. The notion that courts could stop research, however temporarily, did not sit well with a community that was ideologically committed to autonomy in setting the limits of permissible inquiry. Researchers felt that questions of risk had already been answered to the satisfaction of those who best understood biotechnology. Any reanalysis of these issues by judges and lawyers was therefore seen as unproductive, confusing, and damaging to the integrity of science. Bernard Davis, a biologist and outspoken critic of Rifkin's legal tactics, was particularly adamant about the need to restrict risk discussions to specialist forums: "If the courts can distinguish solid scientific evidence from demagogic appeals, fine. But, unfortunately, that is not assured by our present system. To do that, it might be desirable to set up special courts made up of judges who have expertise in science. Without such a sophisticated approach, progress will constantly be held hostage to legal interference by zealots."[4] In a variation on Davis' theme, Daniel Koshland, an editor at *Science,* called for a "judicial impact statement, which would put the onus on the judge or jury that overrules an agency or commission to demonstrate appropriate expertise in the field and an awareness of the consequences of the judgment."[5] Koshland singled out field trials of genetically engineered microbes as an issue on which judges should not rule without first establishing their expertise.

The assumptions of disenchanted scientists like Davis and

Koshland deserve careful probing, since courts in America, for better or worse, will remain an important forum for addressing conflicting public expectations concerning new technologies. The record suggests that despite the fears of many mainstream biologists, legal disputes over biotechnology have resulted more often than not in decisions favoring the interests of science and industry. The most notable instance was the Supreme Court's decision in 1980 to approve the patentability of genetically engineered life forms,[6] but less obvious examples can also be found in cases involving the release of such organisms into the environment. Far from playing an obstructionist role, the courts have helped to normalize genetic engineering by providing forms and methods of discourse that made the applications of the technique seem amenable to control.

The case of biotechnology thus sustains a theme developed in earlier chapters: that the involvement of the courts in science and technology policy often reinforces dominant beliefs and institutional arrangements, which, in this society, include a well-entrenched faith in the progressive force of science and technology. Litigation becomes an avenue for working out, often at an early stage, the compromises necessary for securing social acceptance of a new technology, for example, by making it more responsive to well-articulated social needs, without capitulating to more radical critiques. The courts serve in this way as indispensable (albeit uncomfortable to some) forums for accommodating technological change. For biotechnology, as for nuclear power, litigation served over time mainly to defuse opposition, not to foment it.

The Early Recombinant DNA Debate

Populist lawsuits were far from the contemplation of most molecular biologists when they first broached the question of appropriate controls on genetic engineering, with no prodding from courts or legislatures. The scientists' initiative and its results have been recorded in the science policy literature as an extraordinary, and in some ways strikingly successful, attempt at self-regulation. Subsequent conflicts between legal and scientific institutions over biotechnology can be understood only in the light of this early history, which both framed the issues in particular ways and shaped the expectations of many key actors. A brief summary of the abundant secondary literature on this subject will suffice for our purposes here.[7]

In 1971 a group of cancer researchers became concerned about an experiment designed by Stanford scientist Paul Berg to insert DNA from an animal tumor virus into *Escherichia coli (E. coli)*, a bacterium normally found in the human intestine. The fear was that research along these lines might create a human cancer agent that could escape from the laboratory and, by readily replicating in human organs, produce a public health disaster.[8] Berg responded to his colleagues' concerns by temporarily suspending his experiment. The scientific community continued to discuss the potential hazards of recombinant DNA (rDNA) research at two major conferences held in 1973: the first Asilomar conference, in January, and the Gordon Research Conference on Nucleic Acids, in June. Following these meetings, a committee formed by the National Academy of Sciences recommended in July 1974, in a letter to the editors of *Science* and *Nature* (the "Berg letter"), that a moratorium be declared on rDNA research until adequate controls were developed to guard against any possible risks. The moratorium was generally effective despite increasing competitive pressures on researchers,[9] and the self-regulation movement continued to gather steam in the next few months. At the now famous second Asilomar conference in February 1975, a majority of the scientists in attendance approved a report calling for special safety procedures to be adopted in conducting rDNA experiments.

The task of implementing these recommendations fell almost by default to the National Institutes of Health (NIH), the agency primarily responsible for funding biomedical research, where discussions about regulating work with rDNA had begun even before the Asilomar conference.[10] Originally formed as a "kitchen cabinet" to the NIH director,[11] the Recombinant DNA Advisory Committee (RAC) assumed the task of drawing up guidelines that would allow maximum freedom for research while curtailing any foreseeable risks. As finally issued in June 1976, the NIH guidelines called for two types of "containment," physical and biological. Physically, laboratories were to be classified on a scale from P1 to P4, with extraordinary and expensive precautions (air locks, negative air pressure, special clothing, and so on) recommended for the P4 containment level. Biological containment was to be achieved through use of specially enfeebled organisms, mostly strains of *E. coli,* that would be incapable of surviving outside the laboratory. The legal system, as we shall see, gradually created a third safeguard—the "social containment" that had not been contemplated by the Asilomar scientists.

Periodic relaxation of the guidelines in later years indicated a growing consensus in the research community that early fears about the safety of rDNA techniques in the laboratory had been largely unwarranted. The lay public, too, appeared to fall in line with this assessment. In a 1986 poll conducted for the Office of Technology Assessment by Louis Harris and Associates, 67 percent of those surveyed said that they would either actively favor having a field test site for new organisms in their own communities or would not object to such a site.[12] Widescale use of genetically engineered organisms, however, was opposed by 42 percent, and 42 percent agreed that it would be morally wrong to alter the genetic makeup of human cells.[13] In roughly one decade after Asilomar, public anxiety had apparently shifted away from the laboratory to the social and environmental contexts in which biotechnology would have its commercial applications.

The Limits of Self-Regulation

Although the formation of RAC and the publication of the rDNA guidelines defused public and scientific anxiety, it quickly became apparent that these initiatives were inadequate as a basis for regulating the entire range of technological applications made possible by genetic engineering. The crux of the problem was that the NIH guidelines controlled only laboratory research projects supported by government funds. All other research, particularly that carried out by industry, remained free from control, except when privately funded scientists chose voluntarily to abide by the NIH guidelines.

The limitations of the NIH-RAC regulatory regime could, of course, have been overcome at any time through comprehensive federal legislation. In fact Congress began considering the possibility of such action at the earliest stages of the scientific debate over rDNA research. Senator Edward Kennedy of Massachusetts, a strong proponent of the public's right to regulate biotechnology, and Senator Jacob Javits of New York appealed jointly to President Ford in July 1976 to seek ways of extending the NIH guidelines to industry. Bills proposing to consolidate regulatory authority over biotechnology in the Department of Health, Education and Welfare (now the Department of Health and Human Services) were introduced in both the House and the Senate. The primary effect, however, was to generate industrial and academic opposition to what many saw as unnecessarily stringent controls on a fundamentally safe technology.[14] By the end of the decade, Congress had held many hearings on biotechnology but had lost the momentum to legislate and seemed resigned for the time being to inaction.

During the late 1970s and early 1980s, new issues centering on the commercialization of biotechnology moved to the forefront of public attention. As the first products making use of rDNA techniques were readied for testing and marketing, the absence of comprehensive federal controls again emerged as a serious policy problem, and scientists, industry, and the public all reacted with impatience to the uncertainties of the regulatory environment. The most heated legal, political, and scientific controversy surrounding biotechnology erupted over plans for patenting genetically engineered organisms and for deliberately releasing them into the environment. These disputes drew courts for the first time into the process of policy formulation.

Patenting Life

The U.S. Supreme Court had an opportunity to comment on biotechnology policy as early as 1980 in *Diamond v. Chakrabarty*,[15] a case that has rightly been seen as a milestone in the commercialization of rDNA research. The legal controversy began when Ananda Chakrabarty, a microbiologist at

General Electric, filed an application to patent a bacterium of the genus *Pseudomonas* believed capable of breaking down, and permitting treatment of, multiple components of crude oil spills. Though not created through genetic engineering, Chakrabarty's bacterium (a product of cell fusion) did not previously exist in nature. The Patent Office Board of Appeals rejected the researcher's ownership claim on the ground that the bacteria were living things and hence could not be patented under federal law.[16] The Court of Customs and Patent Appeals reversed, leading the commissioner of patents and trademarks to petition the Supreme Court for reconsideration.

The Supreme Court by a bare majority construed the Patent Act so as to favor Chakrabarty. Section 101 of the law, which defines what inventions are patentable, had originally been drafted by Thomas Jefferson almost two hundred years before the advent of genetic engineering and had remained virtually unchanged since that time.[17] Five justices now concluded that the framer's intent had been expansive enough to encompass all the products of modern human invention, whether living or inanimate. To support this interpretation, the majority cited a 1952 Senate report that recognized as patentable subject matter "anything under the sun that is made by man."[18] The four remaining justices argued unsuccessfully for a more conservative reading of Section 101, holding that so momentous a step as patenting new life forms should not be undertaken without explicit congressional authorization.

In permitting Chakrabarty's bacteria to be patented, the majority pursued a curiously twisted logical path with respect to the relationship between legal and scientific progress. On the one hand, the justices felt that the law should be broadly interpreted so as to serve the changing needs of science and technology, as envisaged by Jefferson's Enlightenment philosophy. On the other hand, Chief Justice Burger advocated a hands-off position on science policy, arguing that the courts were fundamentally powerless to stand in the way of scientific advance: "The large amount of research that has already occurred when no researcher had sure knowledge that patent protection would be available suggests that legislative or judicial fiat as to patentability will not deter the scientific mind from probing into the unknown any more than Canute could command the tides."[19] The law, by this reasoning, was presumed at once to have a duty to foster technological progress and to be powerless to deter or control research. The majority did not attempt to resolve this apparent contradiction between opposition to and acceptance of a theory of autonomous technological development.

Justice Brennan's dissent followed a more coherent strategy in urging the Court to exercise caution when "the composition sought to be patented uniquely implicates matters of public concern."[20] By arguing for restraint in this matter, Brennan displayed a more activist conception of the judicial role than was admitted by the majority. His dissent was premised on a greater faith in the power of patent laws (and, by extension, of judicial decisions construing such laws) to influence scientific inquiry and, presumably, technological progress.[21]

For several years following *Chakrabarty,* the spirit of Brennan's dissent almost seemed to have prevailed over the substance of Burger's majority opinion. The Patent and Trademark Office proved reluctant to capitalize immediately on the Supreme Court's liberal construction of Section 101. It was not until 1987 that the Office signaled, still in somewhat veiled fashion, that it considered multicellular animals to be patentable: it denied a patent request for a genetically engineered oyster[22] but indicated that the reason for denial was the obviousness of the invention, not the inappropriateness of patenting such an organism.

The oyster decision briefly rekindled policy interest on Capitol Hill, mainly in the form of bills proposing a temporary moratorium on patenting higher life forms. A House Judiciary subcommittee began investigating the legal, ethical, and economic implications of such patenting and, in a move that surprised some observers, refused to accept an initial staff report recommending the patenting of animals.[23] In April 1988, however, the Patent Office broke its own unofficial moratorium on the backlog of twenty-one pending patent applications for genetically altered animals. The first such patent was issued to Dr. Philip Leder of Harvard Medical School and Dr. Timothy Stewart of Genentech Inc. for a new transgenic mouse species showing a strong tendency to develop breast cancer. This increased susceptibility made the "Harvard mouse" particularly desirable as a test species for studying the effects of carcinogens and substances thought to protect against cancer.

The Patent Office clearly calculated well in selecting its test case for animal patenting.[24] Although the decision made headlines and further fueled congressional and popular debate, hardly anyone questioned the patent's legality. Acceptance of the status quo finally ratified the conventional wisdom that, in *Chakrabarty,* the Supreme Court had removed all but the economic and psychological barriers to patenting genetically altered animals.[25] In practice, of course, that legal decision only created a rhetorical and normative framing within which other actors (the Patent Office, Harvard scientists, the onco-mouse, and agricultural interests, to name but a few) rearranged themselves over time so as to provide social support for this controversial extension of intellectual property rights.

The Politics of Deliberate Release

For an industry intent on commercializing the fruits of its research and development, the settling of the intellectual property issue in *Chakrabarty* did not provide sufficient policy certainty. For example, biotechnology companies engaged in agricultural research recognized early on that it would be foolhardy to undertake deliberate releases of genetically altered organisms without some kind of official sanction. Yet it was far from obvious which federal agency had the authority to permit such activities. As a research-sponsoring agency, NIH possessed insufficient authority to regulate commercial behavior and could not, in any event, safeguard proprietary information disclosed to it by industry. None of the other federal agencies involved in regulating hazardous products—in particular, the Environmental Protection Agency (EPA), the Food and Drug Administration (FDA), and the U.S. Department of Agriculture (USDA)—possessed a clear statutory mandate to control the products of biotechnological research. The issue of deliberate release thus became embroiled in a complex interplay of forces, as individual agencies sought to expand their bureau-

cratic agendas and the activist community attempted to slow
the process of commercialization.

Several episodes dramatized the confusion and uncertainty
surrounding biotechnology at the beginning of the 1980s. By
late 1984, EPA had reviewed its authority to regulate genetically engineered organisms under two statutes: the Federal Insecticide, Fungicide, and Rodenticide Act (FIFRA), which gov-
erns pesticides; and the Toxic Substances Control Act (TSCA),
governing new chemicals.[26] As a result, EPA had adopted an
interim policy for approving rDNA experiments of a kind not
covered by the NIH guidelines, including the deliberate release
of novel microorganisms. An early test of the EPA procedure
was an application for an experimental use permit by Advanced Genetic Sciences (AGS), a small California-based biotechnology company that wanted to test a genetically altered
microbe—the so-called Ice-Minus bacterium—designed to inhibit frost formation on fruit and vegetable crops.

After about a year of review, including review by its Scientific
Advisory Panel on pesticides, EPA granted AGS a permit to
carry out the intended field test. Within a few weeks, however,
a congressional subcommittee learned from Jeremy Rifkin
that the company had conducted an unauthorized experiment
on the rooftop of its office building. AGS scientists were apparently under the impression that the experiment, which involved the injection of Ice-Minus bacteria into fruit trees, did
not constitute a true "release" and thus was not subject to regulation. Steven Schatzow, then director of the agency's pesticide program, observed, however, that "we all have an idea of
what a contained facility is, I think, and laymen . . . would
agree that a tree does not meet that definition."[27] Interpreting
AGS's action as a deliberate release, EPA temporarily sus-
pended the test;[28] a charge of data falsification and a substantial fine were subsequently dropped or reduced.

At about the same time, Monsanto Company asked EPA to
grant an experimental use permit for a genetically engineered
microbial pesticide for use on corn. The product represented
Monsanto's first major venture into biotechnology and a break
from its earlier exclusive reliance on petrochemicals. By the
time Monsanto requested a permit, it already had a substan-

tial investment in the product: more than three years of research and development, including $250,000 worth of safety
tests in the laboratory.[29] Obtaining EPA's approval was therefore not a matter to be taken lightly.

EPA's nascent regulatory program, however, was not yet prepared for such a challenge. As in the Ice-Minus case, EPA
turned to its Scientific Advisory Panel for a safety evaluation of
Monsanto's product, but the scientists responded with a statement of delphic ambiguity: "In total the data suffice to allow a
judgment to be made to either grant or deny an experimental
use permit. Based on the quality of the data from some studies
the Agency could deny an experimental use permit at this time
and request more information, or the Agency could grant an
experimental use permit on the basis of minimal risk."[30] Expa-
tiating on this rather confusing message, the subpanel observed that some of Monsanto's studies were indeed deficient,
but that the agency could reasonably conclude on the basis of
the remaining studies that the test presented only minimal
risks.

The panel's assessment left EPA in a quandary about how to
proceed. Lacking a clear scientific green light, and still suffering from the disclosure of unauthorized testing by AGS, EPA
decided not to award Monsanto the requested permit. This action forestalled a potentially embarrassing lawsuit, but it also
temporarily halted Monsanto's microbial pesticide program
and created dissatisfaction among scientists who believed that
the pesticide in question was harmless.[31]

The Department of Agriculture was also drawn into controversy over the regulation of biotechnological products. In 1985
USDA's Animal and Plant Health Inspection Service (APHIS)
was asked by Agracetus Corporation, a subsidiary of Cetus
Corporation, to approve field testing of a genetically modified
plant. APHIS determined that the test would not create a plant
pest risk but concluded that it had no authority to give
Agracetus formal permission to proceed. As a result, the company was forced to return to NIH and RAC for approval; the
entire process of obtaining governmental authorization con-
sumed three years.[32]

APHIS also became involved in proceedings to test and license a new genetically engineered vaccine, Omnivac, aimed at controlling an animal disease called pseudorabies. A congressional investigation into USDA's handling of this product revealed pervasive problems in the application of existing law, policy, and procedure:

> Not only did USDA needlessly delay identifying the vaccine as a "recombinant organism" but the principal investigator, Dr. Kit, failed to confirm his interpretation of the NIH Guidelines with the local IBCs [institutional biohazard committees] or with the agencies themselves. USDA's inability to regulate intrastate shipments . . . prior to 1976 was exacerbated by the absence of a clear USDA policy toward the use . . . of data obtained from tests conducted outside the NIH Guidelines for recombinant DNA research. Finally, USDA's tardiness in following its own procedures for review of biotechnology products unnecessarily delayed the marketing of the vaccine after it was proven to be safe.[33]

Frustrated by uncertainty and delay, some companies and researchers tried to bypass the regulatory bottlenecks altogether. For example, in November 1986 Rifkin's Foundation on Economic Trends accused the Wistar Institute of having violated federal guidelines by field testing in Argentina a recombinant rabies vaccine developed with substantial funding from NIH. The NIH director later determined that the charge was unfounded because no federal funds were used for the test itself.[34] In mid-1987 Gary Strobel, a plant pathologist at Montana State University, provoked outrage among regulators and the public by injecting genetically altered bacteria into fourteen elm trees without authorization from EPA, stating that he found the agency's regulations "almost ludicrous."[35] Overwhelmed by the hostile response, a tearful Strobel eventually cut down the trees and was subjected to minor sanctions by EPA and his own university. But many of Strobel's fellow researchers applauded what they regarded as a justifiable act of civil disobedience, a quite different reaction from the universal condemnation aroused by Martin Cline's unauthorized foray into gene therapy (see Chapter 5). Several months later an NIH special committee cleared Strobel, on technical grounds, of charges that he had violated the rDNA guidelines; the committee determined that Strobel's organisms did not fall under NIH's jurisdiction because they did not contain "recombinant DNA molecules" as defined by NIH.[36] The regulations thus played their part in a system of social containment that provided public reassurance while carving out a relatively untrammeled space for research.

Appeals to Law

An environment of confused bureaucratic authority, compounded by scientific and legal uncertainty, provided the ideal conditions for a skilled and ideologically committed pressure group to uncover and challenge shortcomings in the regulatory process. Stanley Abramson, then EPA's associate general counsel, credited Rifkin and his organization with "an uncanny ability to pinpoint weaknesses in our review procedures."[37] It would be more accurate (if less flattering to the regulatory agencies) to say that only a complaisant or blind adversary could have missed those weaknesses in the early 1980s.

The first important lawsuit brought against a federal agency in connection with genetic engineering was not in fact initiated by Jeremy Rifkin. In *Mack v. Califano*,[38] the plaintiffs sought an injunction against rDNA experiments to be carried out at the Frederick Cancer Research Center at Fort Detrick, Maryland, and involving the insertion of DNA from a cancer virus into *E. coli*. The plaintiffs contended that HEW and NIH had not adequately assessed the environmental impact of the challenged experiments pursuant to the National Environmental Policy Act (NEPA). Judge John Lewis Smith of the federal district court for the District of Columbia concluded otherwise. The experiments, he held, clearly would be conducted in accordance with applicable NIH safety requirements, and the environmental impact statement (EIS) already prepared by NIH had taken an adequately "hard look" at the risks entailed by such research. In keeping with the conclusions drawn in Chapter 5, *Mack* displayed the conventional judicial deference toward the laboratory-based activities of scientific researchers.

The NEPA argument resurfaced with greater force in subsequent litigation focusing on deliberate release. In 1984 the Foundation on Economic Trends sought an injunction from the federal district court for the District of Columbia to halt a projected field test of a genetically engineered frost-inhibiting bacterium designed by two University of California researchers, Steven Lindow and Nickolas Panopoulos. FET charged that the defendants (HHS, NIH, and the University of California) had not prepared an environmental impact statement, and they asked the court to enjoin NIH from approving all other such experiments until their environmental consequences had been more completely investigated. In *FET v. Heckler,*[39] Judge John Sirica granted both injunctions. On appeal, Judge J. Skelly Wright, writing for the D.C. Circuit Court of Appeals, upheld the injunction with respect to the Lindow-Panopoulos experiment but did not sustain Judge Sirica's mandate that NIH carry out a programmatic environmental impact assessment for all future releases.[40]

FET's complaint about procedural irregularity was not unfounded. NIH's 1976 guidelines had explicitly prohibited five categories of experiments, including deliberate releases of genetically engineered organisms into the environment. In 1978 a guideline revision authorized the NIH director to make exceptions to these absolute prohibitions but did not provide any criteria for granting such waivers. NIH prepared neither a formal EIS to support this revision nor a less intensive "environmental assessment" to show why an EIS was not required.[41] Accordingly, when NIH began approving field tests in 1981,[42] critics could plausibly argue that the environmental hazards of such experiments had not yet received adequate consideration and that RAC's expertise with respect to such research had not been sufficiently demonstrated.[43]

FET's scientific adversaries, however, viewed the procedural complaint as a mere technicality. They argued that insistence upon an EIS in this case would be an empty formality, since the environmental evaluations carried out by RAC were "functionally equivalent" to an environmental assessment. This counterclaim inevitably led the courts into controversial territory, for they could not settle the question of functional equivalence without assessing the quality of RAC's environmental analysis. As in so many other instances of judicial review, a plausible procedural complaint led into evaluating the merits of a technical dispute that the courts arguably lacked the expertise to understand, let alone resolve.

One factor that particularly troubled both Judge Sirica and the court of appeals was NIH's failure to articulate clear standards for granting waivers to the experiments that the 1976 guidelines had altogether prohibited. The emphasis on standards was consistent with the judicial commitment to "reasoned" administrative decisionmaking (see Chapter 4). In this case, moreover, even the NIH director had explicitly referred to the need for definitive standards in the 1978 guideline revisions and had indicated that the task of developing them should be delegated to RAC.[44] A continued absence of standards in 1984 led a skeptical Judge Sirica to conclude that "the 'standard' for granting a waiver can only be described as whatever it takes to win the confidence of, hopefully, at least a majority of the RAC and the subsequent approval of the Director of NIH."[45] Both Judge Sirica and Judge Wright were on potentially more treacherous ground when they looked into the substance of RAC's environmental analysis. Since the committee had approved the Lindow-Panopoulos experiment by a unanimous vote of 19 to 0, judicial invalidation of the permit entailed an especially conspicuous overruling of expert opinion. Yet, to settle the legally crucial issue of functional equivalence, the reviewing courts had to decide for themselves whether the environmental assessments done by NIH and RAC met the formal standards for compliance with NEPA.

Application of the judicial "hard look" to the RAC's unanimous decision sent warning signals to the presiding judge to which few scientists would have been as sensitive. One significant environmental issue raised in connection with the Lindow-Panopoulos experiment was the possibility that the genetically engineered organisms might disperse into the environment. In reviewing RAC's treatment of this issue, the court discovered, first, that the committee's conclusion was only one sentence long.[46] Second, even this single sentence was taken almost verbatim from the Lindow-Panopoulos proposal, suggesting that there had been no independent analysis by the committee. Third, there was no explicit discussion of possible consequences if a very small number of test organisms in fact migrated beyond the periphery of the treatment locations. Given these facts—all calculated to undermine credibility in a framework of legal analysis and discourse—the court predictably concluded that NIH and RAC had completely failed to address a major environmental concern. Perhaps just as predictably, the scientific community failed to concur in this judgment.

Conflicting Interpretations

Maxine Singer, a prominent biologist known for her enlightened role in organizing the first genetic engineering moratorium, represents what may well have been the contemporary scientific consensus on *Heckler*. Deploring the gap between the cultures of law and science, Singer strongly criticized Judge Sirica for overruling the RAC:

> there was no reason whatever to believe that the proposed field test would affect the human environment at all, let alone significantly. Moreover, it is difficult to think that seeding a tiny potato patch with a relatively incompetent variant of a common bacteria is a major federal action. The plaintiff's arguments that were accepted by Judge Sirica may make sense from a narrow legal perspective, but they make little sense in the context of available scientific knowledge. I cannot help but wonder whether the case might not have been found frivolous to begin with, if the Judge had understood the science.[47]

Singer thus aligned herself with fellow scientists Davis and Koshland in suggesting that disputes over biotechnology should be decided only by judges who "understand the science."

In its own way, however, Singer's analysis of Judge Sirica's opinion misunderstood the law at least as deeply as she believed the judge to have misunderstood the science. Singer's error was to equate scientific legitimacy with legal legitimacy. According to her view of the situation, the court should have ratified the Lindow-Panopoulos experiment because it had undergone "rigorous scrutiny by the NIH advisory committee charged with oversight of recombinant DNA experiments."[48] This call for deference to expertise overlooked the fact that public decisionmaking on science and technology must conform to additional standards of legitimacy that cannot be derived from technical expertise alone: adherence to the letter and spirit of the law, reasoned explanation of official actions, and an administrative record that satisfies the public that decisions do not reflect merely the narrow self-interest of scientists.

Singer's refusal to see how "seeding a tiny potato patch" could constitute a "major federal action" illustrates a scientist's perhaps understandable propensity to read physical, or naturalized, meanings into legal language, a practice that not surprisingly may lead to serious misinterpretations of the law.[49] The federal action at issue in *Heckler* was not the experiment itself, with its significance defined by its physical scope, but the 1978 guideline change that relaxed the absolute prohibition on field tests and empowered NIH to approve such tests on a case-by-case basis. Hence, neither the small size of the potato patch nor the survival capacity of the frost-preventive bacteria was relevant to determining whether these policy changes were "major." Instead, the court quite properly looked to the NIH director's own declaration that the guideline change *was* a major federal action, a declaration that the federal defendants never bothered to contest.[50]

But should the courts nonetheless have dismissed the suit as raising only a harmless or insignificant procedural error? Since a properly constituted technical authority, the RAC, had approved the experiment's safety and validity, was it appropriate for a lay court to overrule the decision on purely procedural grounds? Given the unanimity of the RAC vote on frost-preventive bacteria, the court could without difficulty have regarded the procedural defects alleged by FET as trivial; in other words, it could have concluded that additional procedures would not have changed the outcome. In administrative law, this is ordinarily a good ground for *not* overturning an agency decision. Singer and other scientists were probably right to think that scientifically trained judges would have dismissed the absence of an EIS as a mere technicality and accepted RAC's determination as final.

Legal writers, by contrast, approached the biotechnology suits from a perspective that emphasized the forms of law over the substance of science. Hallmarks of the legal mindset include a concern for process and a predisposition to see judicial review as a panacea against abuses of administrative discretion. Thus, a 1986 commentary on *Heckler* attributed NIH's deficient environmental analysis to tunnel vision and suggested that "this institutional shortcoming would have been minimized if decisionmaking had been subjected to appeal and review before an interdisciplinary body."[51] Another article criticized *FET v. Thomas,* a decision upholding EPA's permission to field test a frost-preventive bacterium, as being overly deferential to the agency. The proposed antidote was for Congress to prescribe a stricter standard of judicial review.[52] The article also suggested that EPA's decisions should be scrutinized prior to judicial review by a multidisciplinary expert committee in order to ensure that they are technically sound and reflective of ecological and popular interests.

Inclined to be skeptical about the integrity of federal regulators, as well as about their technical and policy expertise, these legal analysts favored activist judicial review as a means of holding administrators accountable. Such proposals are more interesting for the insights they provide into legal culture than for their practical wisdom. A new standard of review not only would have been excessively legalistic, but very likely would not have led to substantially different outcomes for the biotechnology industry.[53] Adding another layer of review would in any case have foundered against the demands for cost-cutting and bureaucratic streamlining that resurfaced in U.S. regulatory policy in the 1980s and beyond.[54]

Retreat from Conflict

In retrospect, *Heckler*'s noisy reception was entirely out of proportion to its ultimately slight impact on the regulation of biotechnology. Following *Heckler,* FET seemed destined to lose many more cases than it won. In 1985 FET successfully sued the Department of Defense to block construction of a proposed testing facility for chemical and biological warfare agents on the ground that environmental impacts had not been adequately addressed.[55] A year later, however, in *FET v. Thomas*,[56] the Foundation lost its bid to prevent EPA from authorizing an outdoor test of frost-preventive bacteria. In *FET v. Johnson*,[57] the D.C. district court denied a challenge to the Coordinated Framework for Regulation of Biotechnology because the plaintiff lacked standing. FET had failed to demonstrate any specific injury or likelihood of immediate hardship. And in *FET v. Lyng*,[58] the D.C. Circuit ruled that USDA was not required to prepare a programmatic EIS in support of its animal productivity research program. Understandably perhaps, Rifkin's widely publicized campaign against the marketing of genetically engineered bovine growth hormone in the early 1990s made much more use of direct-action strategies, such as supermarket boycotts, than of legal ones.

By 1984 as well the White House under President Reagan had begun to regain control over biotechnology policy. In December 1984 a Working Group on Biotechnology, operating through the Office of Science and Technology Policy (OSTP), published a "Proposal for a Coordinated Framework for Regulation of Biotechnology."[59] The document included policy statements by EPA, FDA, and USDA about their plans for accommodating biotechnological products and processes within their existing regulatory programs. Following public comment, an interagency group entitled the Biotechnology Science Coordinating Committee, again working under OSTP's auspices, published in the *Federal Register* a revised and expanded "Coordinated Framework for Regulation of Biotechnology."[60] The new document incorporated policy statements from NIH and OSHA along with statements from the three agencies participating in the 1984 proposal.

The final version of the Coordinated Framework formalized an interagency consensus that the problems encountered in regulating biotechnology could be adequately handled under existing statutory authority, although participants agreed that some new regulatory initiatives would be required from individual agencies.[61] Congress would not need to act because, as participants in the Framework agreed, biotechnology as a *process* presented no risks that were novel enough to require the legislature's attention. Rather, it was the individual *products* of biotechnology that had to be evaluated for safety, and this evaluation was perfectly feasible under the mosaic of regulatory statutes that Congress had put in place during the 1970s. An important and perhaps not unintended consequence of this decision was to remove the courts from the foreground of policy formulation. Historical experience suggests that a new federal law on biotechnology would inevitably have opened up new spaces for interpretive conflict and hence new grounds for litigation; by contrast, making policy through administrative regulation under older laws circumscribed the potential for legal activism.

Evidence that the social debate about the release of genetically engineered organisms has been effectively contained comes from many quarters. Although disagreements of regulatory philosophy and approach remain among the federal agencies, these are apt to be flagged and deliberated on the pages of technical and professional journals rather than in more open, lay forums.[62] Several states have enacted laws specifically aimed at biotechnology in order to create a more stable environment for industry than exists under the federal government's Coordinated Framework. This legislation more firmly institutionalizes the federal regulatory approach, but without reopening national debate.[63] Finally, comparisons with several European countries show that the public debate on biotechnology in the United States neither unpacked the social and moral dimensions of genetic engineering as completely nor resulted in as stringent regulatory controls as in industrial nations on the other side of the Atlantic.[64]

A Disciplined Debate

Scientists' objections to *Heckler,* the only significant federal court decision that challenged rDNA research, rested on a bedrock conviction that lay persons should never have the last word in determining when experimentation should go forward.

But placing the biotechnology lawsuits of the 1980s within a longer political and historical narrative undercuts any suggestion that the courts have seriously impeded the progress of this emerging technology. The administrative disarray and legal confusion surrounding the early efforts to regulate biotechnology generated a variety of disputes that were legitimately the concern of the courts. While federal judges at times estimated the physical risks of rDNA research differently from biologists, their actions partly reflected the scientific community's own inability to communicate a convincing message about safety. Moreover, the flurry of litigation in the 1980s left intact a risk-based, heavily bureaucratic approach to regulating biotechnology that did not in the end encourage open-ended moral or ethical questions.

The Supreme Court's momentous verdict on the patentability of genetically engineered animals is a useful reminder that permission from the law is not alone sufficient to elucidate or to resolve complex questions of a technology's social acceptability. *Chakrabarty* gave the Patent Office unlimited legal authority to issue animal patents, but the eight-year lag before this authority was officially invoked and the congressional outcry following the patenting of the "Harvard mouse" show that many more elements were involved in creating new property rights in this area than a technically "correct" interpretation of patent law. The Supreme Court's narrowly legalistic construction of the patent question probably inhibited the kind of broad public debate that would properly have accompanied policy development on this morally and ethically charged issue.

The FET initiatives, by contrast, sought to generate a level of discussion more nearly in keeping with the wide-ranging potential impacts of biotechnology. That FET achieved so little success in the long run attests to the limitations of legal proceedings as a forum for framing and conducting meaningful technology assessment. It is instructive, in this connection, to introduce an outsider's perspective on the alleged divisiveness of the U.S. biotechnology lawsuits of the 1980s. Edward Yoxen, a British expert on biotechnology, interpreted the U.S. controversies in the following light:

in the United States, public debate, some of it caricature, fantasy and phobia from "both" sides, and some of it highly informed and rational, has served to get implicit assumptions scrutinised and dogmatic beliefs revealed as such. America comes out of all this as a society more committed to, and much more at ease with, adversarial debate, and more used to staging conflicts as a means to their resolution rather than pretending they are not there.[65]

If, as Yoxen suggests, America's adversarial style of debate leads to more comprehensive and stable outcomes than in other countries, then advocates of technology should think twice before encumbering our litigation-centered policy process with expensive new add-ons, such as Bernard Davis' specialized tribunals or the multidisciplinary review committees favored by some lawyers. The relatively narrow and possibly premature "consensus" created by such bodies would be very likely to unravel in the face of an unexpected disaster or new scientific findings about risk. But Yoxen's reference to "staging conflicts as a means to their resolution" leaves us with more disturbing thoughts. From the standpoint of biotechnology's progress, it is tempting to see both *Chakrabarty* and *Heckler* in retrospect as "staged" conflicts indeed: rituals that, in the formal guise of conflict and resolution, opened only a limited and technical space for dissent, and then only to close it the more firmly with the magisterial authority of the law.

8

Family Affairs

As we saw in the last two chapters, there is not much question in America's technological culture that the state has primary responsibility for controlling risks to public health, safety, and the environment. These hazards are thought to present a problem of known, or in principle knowable, boundaries and of manageable character, although there may be disagreements about the magnitude of a particular risk (for instance, about how many cancers, what likelihood of gene transfer, how much projected exposure to a pollutant to expect in a given situation). Conflicts about health, safety, and environmental risks are therefore resolved within a conceptual framework whose basic elements are already fixed in ways that do not have to be renegotiated from case to case. Claims about new risks, when a court chooses to recognize them at all, can be analogized with relative ease to old ones.

By contrast, the initial framing of the risks created by reproductive and life-prolonging technologies takes place in less settled territory. Are we dealing here with a moral problem, a technological one, or, more prosaically, one of limited access and resources? Is it a new social problem or one that can be handled by existing arrangements in society without significant legal and policy innovation? The issues of who should control these technologies, and by what means, are therefore less clear-cut; plausible claims can be made on be-

160

half of individuals, professional organizations, interest groups, and the state. In this and the next chapter we will assess how effectively the courts have performed in first framing and then resolving the conflicts of governance and control that arise from changing capabilities in biomedical science and technology.

The controversies addressed in this chapter all center in one way or another on the legal status of the human fetus within a changing constellation of public expectations about sexuality, women's liberty, childbearing, parenthood, and the family. The human fetus, until recently, was a largely invisible and voiceless member of society. Technological innovations over the past few decades have given the fetus greater physical reality and new claims to legal rights while at the same time offering women more grounds for preventing, redefining, and even terminating pregnancy.[1] Conflicts associated with expanded technological options for contraception and abortion offer one vantage point on these issues. Another set of disputes concerns the gradual uncoupling of biological reproduction from social parenting through technological means such as artificial insemination, *in vitro* fertilization (IVF), and embryo implantation. Intersecting with the reconfigurations of the family through adoption and divorce, these unconventional reproductive pathways have begun to undermine the accepted meanings of "mother," "father," "child," and "family."

Technological developments around reproduction have opened up areas of interpretive flexibility around issues that were previously thought to be settled (or, in the terminology of the social construction of technology, "closed" or "black-boxed").[2] Social taboos once seen as unyielding are no longer believed to be so, and new social choices are revealed in areas that formerly were ruled exclusively by biology. The law is called upon to fix the leaky walls between the worlds we construe as social and natural, to recreate normative order when previous understandings have been stretched to the fraying point. How should the rights of pregnant women be balanced against those of their unborn children, and should it matter whether the fetus is viable or could be made viable through medical technology? Should a doctor be held liable for failing to administer a diagnostic test, thereby "causing" the birth of a defective child? Who "owns" a frozen embryo when the biological "parents" are dead or divorced, and do property rights provide the most compelling framework for resolving these issues? Should a woman have a right to be impregnated by sperm from a husband to whom she is no longer married? And how should women's conduct be regulated when they agree, for love or money, to bear children for others to nurture?

Perhaps more obviously than in other areas of technological change, biomedical technologies redraw existing boundaries between the natural and social worlds and reshape people's expectations of agency and control. In assessing judicial responses to these technologies, we will ask, consistent with our interest in deconstruction and civic education, how clearly courts have discerned these phenomena and articulated them through the "literary technology" of legal opinions.[3] We will also probe the capacity of courts to act as critics of their own role in creating new norms related to technology, and evaluate how effectively they have translated their understandings into workable rules of conduct.

Making the Meaning of "Privacy"

Controversies over the wisdom, morality, and constitutionality of the Supreme Court's 1973 decision in *Roe v. Wade*[4] injected a peculiarly strident note into American political discourse for more than twenty years. One's stance on abortion became for a time a reliable indicator of one's position on a range of other social issues of the day. Under Presidents Reagan and Bush, beliefs about abortion were used as a "litmus test" for deciding who should hold high public office, from federal judgeships to cabinet posts, and even the directorship of the National Institutes of Health. Not until Justice Ruth Bader Ginsburg's confirmation hearings in the summer of 1993 did strong bipartisan support develop for a Supreme Court nominee who openly championed liberal abortion rights.

Summarizing the voluminous scholarly literature on *Roe*[5] would not greatly advance the purposes of this book: our concern here is not so much with whether the case was rightly or wrongly decided as with its role in framing and stabilizing (or at times failing to stabilize) a technologically mediated revolution in human sexual behavior. In reconsidering *Roe* from this perspective, we may usefully draw upon the concept of the "actor network" in the sociology of technology. Complex technological systems have been shown to consist of a network of heterogeneous actors, all of whom must coexist in a mutually sustaining and disciplining relationship to make the technology work. Elements of such a network include inanimate objects, measurement techniques and protocols, instruments, trained human actors, and social institutions, including the law.[6] If these components do not operate in harmony, the technological system falls apart. A successful constitutional principle (analogous to a key scientific discovery or an autonomous machine) seems capable on the surface of holding a complex social and behavioral system together by dint of a stated normative standard. In practice, the principle may not "take" once society begins to interpret it, and may indeed unravel to the point of being discarded, unless and until the remaining elements of the social network are prepared to be held in place by it. Pathbreaking constitutional decisions such as *Roe v. Wade* are vulnerable not only because they are subject to the law's internal deconstructive mechanisms (legal criticism and testing), but also because the social network whose behavior they hope to influence proves too resistant.

David Garrow's monumental study of the making of *Roe* provides insights on its construction as a piece of literary technology.[7] Garrow describes in painstaking detail the negotiations that took place within the Court to find appropriate framing concepts for abortion rights, as well as the reasoning to support them. Three landmark cases spanning nearly thirty years—*Griswold v. Connecticut*,[8] *Roe v. Wade*, and *Planned Parenthood of Southeastern Pennsylvania v. Casey*[9]—illustrate the difficulties the Court faced in fashioning language that would be firm yet flexible enough to sustain all the pressures placed on it. To do this, the Court needed to define and delimit the freedom afforded to women by modern methods of contraception; to resolve the contradictions that this freedom created in relation to the rights of the fetus and the state; and, not least, to articulate its conclusions in a way that would command assent from an unruly conglomerate of actors such as Congress, state legislatures, activists for and against abortion, physicians, researchers, and, not least, critical legal academics.

The issue of women's reproductive freedom came to the Supreme Court through challenges to an 1879 Connecticut law that criminalized the use and prescription of contraceptives. Justice William O. Douglas, an ardent if unscholarly defender of individual liberties, wrote the brief majority opinion in *Griswold v. Connecticut,* announcing that decisions regarding contraception are made within a "zone of privacy" that should be insulated from state intervention: "Would we allow the police to search the sacred precincts of the marital bedroom for telltale signs of the use of contraceptives? The very idea is repulsive to the notions of privacy surrounding the marriage relationship."[10] It seemed at one level an easy decision: even the two dissenting Justices, Hugo Black and Potter Stewart, were persuaded that the Connecticut law intruded too far into decisions that should be made in private by marital partners and their physicians. Yet the six separate opinions in *Griswold* told a more complicated story. The justices clearly found it hard to agree on a single rationalization of a social insight that struck all or most of them as straightforward. Somewhat reluctantly, the majority coalesced around the conclusion that it was "privacy" that the state law had violated.

But *Griswold* left painfully ambiguous the precise meaning of the privacy right that it purported to find in the Constitution. Was it the right to conduct certain intimate relationships free from governmental control (a right the Court explicitly refused to extend to homosexual couples in *Bowers v. Hardwick*);[11] was it a right available only to married couples (but then extended without conflict to unmarried persons in *Eisenstadt v. Baird*);[12] or was it merely the right to buy and use pharmaceuticals without hindrance by the state (a position that Laurence Tribe felt had to be explicitly disavowed when he was arguing to the Court years later in *Hardwick*)?[13] And where in any case did the right to privacy come from? Was it an "emanation" from the "penumbras" of other, more explicit constitutional guarantees, as Douglas wrote in his majority opinion? Could it be more solidly grounded, as Justice Arthur Goldberg argued, in the Ninth Amendment, which left residual powers to the people? Or would the Court have done better to follow Justice John Harlan and to use the concept of "liberty," a word explicitly mentioned in the Constitution? *Griswold* in this respect failed to close the domain of interpretive flexibility around the idea of privacy through a compelling use of constitutional language.

The never fully legitimated concept of privacy was pulled into yet more contested territory when the proponents of reproductive freedom brought to the Supreme Court their challenges to state antiabortion laws. Now the Court had to balance women's reproductive autonomy against the rights of a gestating fetus that would at some point take on the full attributes of a human being.

By 1973 new contraceptive technologies had not only made it possible for women to plan their pregnancies with great predictability but had also empowered both women and men to engage in sexual activity without intending to create a family. Sexual involvement no longer entailed an almost certain loss of liberty, a change that was especially important for women, the biological childbearers, whose social status had historically been defined by their inability to control the timing of reproduction and motherhood. The availability of contraception and its ready acceptance by major social institutions (the Catholic church was an exception)[14] created a more open moral space for abortion by separating sexual relations from reproduction. But within this space, restrictions on abortion undercut precisely the freedoms that contraception held forth, since unwanted pregnancies could still result from "unprotected sex"— and against this consequence the law provided no safeguards. Pressed on one side by the exhilarating prospect of liberating women from biologically defined gender roles and, on the other, by older views of womanhood and the still uncertain moral status of the fetus, especially at the early stages of gestation, the differential access to contraception and abortion became increasingly the focus of legal and social strife.

To recognize, as many did by 1973, that there was a morally

indeterminate continuum from contraception to early abortions constituted at best a naked social insight that still had to be clothed in persuasive doctrinal language. Here the *Roe* Court stood on shaky ground, constrained by its own precedents. Contraceptive use had plausibly been represented as "private" in *Griswold* and succeeding decisions because it was an activity that few saw merit in constraining. *Roe,* however, brought into the nexus of women, physicians, and the state a new actor, the fetus, whose interests could not easily be contained within the marital bedroom, which *Griswold* had constructed with some difficulty as the relevant space for exercising sexual freedom. The *Roe* majority attempted to recognize the rights of the fetus by balancing the conflicting interests of women, unborn children, and the state. The result was the controversial trimester framework, which offered something to all concerned parties but on a rationale that crumbled in the face of political opposition. *Roe*'s most significant contribution was to recognize as a matter of constitutional principle that women have a right not to carry each and every pregnancy to term. More controversially, *Roe* held that this right, which the Court again described as a right to privacy, should be absolutely protected during the first trimester of pregnancy. In the second trimester, an increased risk of death from abortion justified, in the Court's view, state regulation to protect maternal health. Only in the third trimester, past the point at which the fetus could survive outside the mother's womb, did the balance finally tilt toward the protection of fetal life, and states were left free to restrict or prohibit access to abortion.[15]

The trimester approach addressed a moral dilemma through what looked suspiciously like a balancing of political interests, but the interests at stake were unequally represented in the Court's deliberations. *Roe*'s central concern with the limits of women's sexual freedom had, of course, been percolating through the legal system for many years. Accordingly, for this aspect of the decision, there were grounds for Garrow to conclude that "in the end, Harry Blackmun grasped perhaps the simplest but eventually the most long-forgotten truth of all: '*Roe* against *Wade* was not such a revolutionary opinion at the time.'"[16] But the resources that the Court employed to delineate the trimester framework, such as Blackmun's ad hoc history of abortion laws and deference to medical opinion, proved vulnerable. In particular, the artificiality of the boundary between the first and second trimesters quickly became apparent as improvements in medicine made virtually all second-semester abortions safe from the standpoint of maternal health. Justice Sandra Day O'Connor called attention to this point in her dissenting opinion in *City of Akron v. Akron Center for Reproductive Health, Inc.*, predicting that medical technology would move forward the point at which states could justifiably regulate in the interests of maternal health, but move backward the point of permissible regulation in the interests of the fetus. These developments, she suggested, would put the trimester scheme on a collision course with itself. Courts then would have to "pretend to act as science review boards,"[17] since the constitutionality of state abortion laws would depend on case-by-case determinations of the capabilities of medical technology.

Later medical advances failed to support O'Connor's speculation that viability, in the form of technologically assisted survival, would be pushed indefinitely far back in pregnancy.[18] Indeed, as a load-bearing legal concept, viability acquired greater importance over time precisely because it did not yield to technological advances. Garrow notes that Blackmun initially picked viability as the appropriate point for state intervention because he was persuaded by Justice Thurgood Marshall that any earlier point would possibly undermine the right secured to women by *Roe*.[19] By coincidence, newer medical knowledge brightened the viability line as a suitable point for articulating fetal rights just as surely as it dimmed the first trimester as a meaningful boundary for the rights of women. The Court gradually backed away from *Roe*'s philosophically questionable position that the state's interest in potential human life could be expressed only after the point of viability. In *Webster v. Reproductive Health Services*,[20] the Court recognized the rights of state legislatures to protect fetal life throughout pregnancy. *Casey* completed this evolution: here the majority explicitly rejected the trimester framework and held that states could regulate abortions at any stage provided that their actions did not "unduly burden" a woman's fundamental right to terminate her pregnancy.[21]

The fires of conflict lit by *Roe* did not appear likely to be put out completely by *Casey*'s "unduly burdensome" standard, which seemed on its face to require detailed, case-by-case inquiry into the social impacts of particular policies, including technology-dependent policies such as mandatory viability testing.[22] The increasingly violent tactics of antiabortionists, such as the murder of two Planned Parenthood workers at clinics in Brookline, Massachusetts, in December 1994, demonstrated the movement's continued political resolve, even if now increasingly directed outside the law. Another indicator of *Roe*'s imperfect ability to discipline technological change was the political history of RU-486, the so-called morning-after pill, whose importation from France was successfully prohibited by the Department of Health and Human Services during the Bush administration.[23] If the moral acceptability of early abortions had been more securely established, the administrative ban on RU-486 would have proved legally difficult to sustain. As it is, President Clinton's widely approved repeal of the ban in January 1993 showed that political support for liberalization of birth control options had solidified in spite of continued uncertainties in abortion law.

Nonetheless, the often-repeated assertion that *Roe* was "wrongly reasoned" conveys a flat and unduly negative appraisal of the decision's contributions to social and technological change. It misses the point, for example, that the Supreme Court in 1973 correctly perceived that birth control and sexual liberation had begun to place stresses on the moral boundary between contraception and pre-viability abortions, a boundary that put too much weight on whether action to "prevent" pregnancy was undertaken shortly before or shortly after the sex act. *Roe,* as narrowed and reaffirmed in *Casey,* also pointed ahead to the fact that, in a world of controllable reproduction, "viability" rather than conception would prove to be for most women and men the morally salient boundary in making abortion decisions. Significantly, in the twenty years between *Roe* and *Casey* the Court also appeared to have gained new insights into the interpenetration of constitutional reasoning and social action. In a display of symbolism and resolve that had been missing in earlier decisions, the majority opinion in *Casey* was signed by Justices Sandra Day O'Connor, Anthony Kennedy, and David Souter and thus became the first decision in twenty-three years to bear the name of more than one author.[24] Souter's contribution to the opinion spoke eloquently of women's changing status and noted, in a rare reflexive turn, that *Roe* had played an integral part in the way people had for two decades "organized intimate relationships and made choices that define their views of themselves and their places in society."[25] As literary technology, then, *Roe* finally demonstrated its persuasive power through its incorporation into the majority opinion in *Casey*.

Mapping the Domain of Fetal Rights

While abortion politics absorbed seemingly inexhaustible supplies of legal energy on the national scene, lower courts were confronted with scores of decisions that were peripheral to the main controversy but were nonetheless critical in redefining public understandings of reproductive freedom in an age of rapid biomedical discovery. As in the abortion cases, many of these conflicts centered on the status and entitlements of the developing fetus. Courts across the board performed better at providing ad hoc resolutions for localized troubles than at crafting a coherent vision of changing biological and social relations. Collectively, however, these cases no less than the abortion decisions confirmed a growing public consensus that different views of fetal rights legitimately come into play in conflicts at different stages of fetal development.

Embryonic Conflicts

The status of medically created embryos was one issue for litigation. The federal government's reluctance to take proactive positions on human reproduction relegated policymaking on new reproductive techniques almost entirely to state institutions, just as it left research and development largely to the private sector. In 1979 the Ethics Advisory Board (EAB) of the (then) Department of Health, Education and Welfare issued a report on policies for *in vitro* fertilization and recommended increased federal support for research.[26] But these stirrings of interest ended with the EAB's dissolution in the Reagan administration, a step that foreclosed possibilities for a deliberative exploration of the medical, legal, and ethical dimensions of IVF, let alone for coordinated policymaking at the federal level.[27] Federal funding for IVF research, too, was placed under an indefinite moratorium.[28]

Through the 1980s techniques such as egg donation, IVF, embryo freezing, and embryo transfer advanced rapidly, creating science-fiction-like scenarios in which the human embryo appeared as an unexpected actor. In the absence of national policy, conflicts over the embryo's status were regulated under a patchwork of state laws, some explicitly and some only coincidentally applicable to IVF. Lawsuits aimed generally either to constrain or to increase people's access to technologically assisted reproduction. Thus, in *Smith v. Hartigan*,[29] one of the few recorded cases relating specifically to *in vitro* fertilization, a childless couple and their physician challenged the constitutionality of an Illinois statute that appeared to prohibit access to IVF.

Embryos clearly were not treated as persons in their own right in *Del Zio v. Presbyterian Hospital in the City of New York*,[30] one of the earliest cases to test the legal implications of IVF. Mrs. Del Zio had decided, in consultation with her personal physician, to undergo IVF at a time when the procedure was still regarded as experimental. A culture prepared with ova from Mrs. Del Zio and semen from her husband was placed in an incubator at Columbia University's Presbyterian Hospital, where it was supposed to remain for four days. On the second day Dr. Vande Wiele, then chairman of Columbia's Department of Obstetrics and Gynecology, had the culture removed to a deep freeze, effectively terminating the procedure. Mrs. Del Zio thereupon sued Presbyterian Hospital, Columbia, and Dr. Vande Wiele for theft of personal property and emotional distress. An eventual award of $50,000 in damages not only expressed sympathy for Mrs. Del Zio's lost reproductive expectations but also implicitly ratified her claim to ownership of the destroyed embryos.

In *Davis v. Davis*,[31] by contrast, a woman in Tennessee claimed a legal right to be implanted with frozen embryos against the wishes of a now divorced husband. The court of appeals held that a decision favoring the "mother" in this case would violate her ex-husband's right not to procreate. The embryos themselves were not entitled to be protected as "persons" under Tennessee law, which in this respect was consistent with *Roe*. Yet, unable to put either biological partner's interests above the other's, the court awarded them joint custody over the now childlike embryos. In contemporaneous cases involving artificial insemination, a single claimant's right to have children with sperm from a deceased partner provoked no comparable analysis, even though the child born from such a procedure sometimes faced unforeseen legal uncertainties.[32] Wishing to assert the embryo's personhood, antiabortion groups in Illinois successfully obtained legislation that gave legal custody of a fetus produced through IVF to the physician who performed the procedure.[33] Ironically, the defenders of the law claimed that a medical decision not to implant an IVF embryo could be regarded as a lawful termination of pregnancy— that is, as a form of abortion—hoping by this means to avoid conflicts with *Roe v. Wade*.

Imperfect Lives

The convergence between new capabilities in genetic testing and the liberalization of abortion under *Roe v. Wade* created another locus for renegotiating the value and meaning of prenatal life. Only a generation ago, the birth of a physically or mentally impaired infant would have been regarded as a misfortune decreed by the will of God. Today chromosomal abnormalities such as Down's syndrome, genetic diseases such as Tay-Sachs, and neural tube defects such as spina bifida are detectable through amniocentesis, a procedure that uses fluid drawn from the amniotic cavity at about the sixteenth week of pregnancy. A newer test, chorionic villi sampling, makes use of tissue from the developing placenta to detect many of the same abnormalities as amniocentesis.[34] Together with related techniques such as sonograms, ultrasound, and blood tests, these procedures can provide women in high-risk pregnancies with extremely accurate information about the future physical condition of their unborn babies. Medical treatment, however, remains beyond reach for most hereditary defects picked up through prenatal testing.[35] Accordingly, the only realistic way to prevent the birth of a severely diseased or deformed child is to seek a lawful abortion. Parents, convinced by testing technology that children have a "right" to be born free of certain congenital defects,[36] are prepared to sue their physicians for failing to warn them in time of the risks of such a birth.

U.S. courts have recognized at least since 1946 that a child born alive is entitled to damages for injuries caused by prenatal medical malpractice.[37] In early cases of this type, however, the physician's negligence had either produced or exacerbated the claimant's condition. In a 1967 case called *Gleitman v. Cosgrove*,[38] the New Jersey Supreme Court was asked to extend malpractice principles to a claim arising from an alleged failure of prenatal counseling. Mrs. Gleitman's physician had not warned her that rubella (German measles) contracted early in her pregnancy might harm the fetus. Following the baby's birth, Mrs. Gleitman argued that the doctor's inaction had deprived her of the opportunity to choose a timely abortion. The court held that these facts did not establish a claim for compensating either the child or the parents.

Following *Gleitman*, courts slowly disentangled two separate, though closely related, torts related to negligence in prenatal counseling: a "wrongful birth" claim[39] by parents against a physician who failed to inform them of the risk that their child would be born with birth defects; and a "wrongful life" claim by the child, who would not have experienced a life of suffering if the parents had been properly advised. "Wrongful birth" claims represented a relatively straightforward extension of traditional negligence law. Where there was evidence of inadequate testing or counseling, courts had little difficulty holding that the physician had violated a duty of professional care to the parents and that this breach of duty was the proximate cause of the birth of the deformed child.[40] Parents could then be compensated for unusual expenses associated with rearing a mentally or physically impaired child.[41] "Wrongful life" claims presented more formidable analytic challenges.[42] How could a court calculate the damages due to the disabled child when common law judges were used to treating the value of any life, even an impaired one, as immeasurably greater than that of nonlife? The *Gleitman* court indeed asserted that measuring damages in a wrongful-life action was a "logical impossibility,"[43] and in *Berman v. Allan*[44] the New Jersey Supreme Court stated that existence, even with deformities, was preferable to nonexistence. Yet, seventeen years after *Gleitman*, in *Procanik v. Cillo*,[45] New Jersey followed California and Washington in granting a newborn plaintiff the right to collect some damages for injuries resulting from an almost identical breakdown in prenatal counseling.[46]

Sorting out the basis for liability, however, opened the way to wider social and cultural issues that courts could not as competently resolve. Wrongful-birth and wrongful-life actions came to be seen as another judicially sanctioned right that established impossibly stringent standards for physicians. The result, critics complained, was to encourage the practice of "defensive medicine," with its trailing array of expensive and unnecessary tests.[47] That physicians who specialize in prenatal care should be held to the established standards of their profession was not in doubt.[48] There was little disagreement, for example, that courts could reasonably impose liability for failing to ask about a family history of birth defects, for withholding information about amniocentesis from older parents, or for carelessly performing prenatal diagnostic tests.[49] The problem with new theories of liability, however, was that they put unacceptable pressure on the boundaries of established medical practice; even small changes in liability rules created sufficient legal uncertainty to push medical diagnostic strategies beyond limits set by professional practice.

Physicians and their organizations are complicit in using technology as a safeguard against possible legal assaults on their judgment. George Annas, the well-known expert on health and family law, describes an occasion when the American College of Obstetrics and Gynecology (ACOG) advised its members to inform every prenatal patient that an alpha-fetoprotein test was available for screening against neural tube defects.[50] This advice, which was calculated to increase the demand for tests, ran counter to the opinion of ACOG's own medical staff that screening all pregnant women was of uncertain value, since the test could not always be coordinated with adequate counseling and technical support. Annas concluded that ACOG's rationale for its change of policy was "not medical, but legal: to give the physician 'the best possible defense' in a medical malpractice suit."[51] Put differently, liability law and medical practice acted synergistically to remove discretion from people and invest it instead in the impersonal technology of the "test"—seen in our culture as objective and untouched by judgment.

In a society that is strongly attracted to the ideal of physical perfection, wrongful-birth and wrongful-life actions also foster the impression that every child has a right to be born whole[52] and, conversely, that the birth of damaged children should be prevented whenever possible. As testing techniques advance, the legal system may be under greater pressure to decide which children are "normal" and what kinds of existence are "acceptable." In this way, wrongful-birth and wrongful-life actions could become the instrument of a new eugenics. Prodded by the threat of liability, physicians may well collaborate with families to eliminate characteristics that society at present considers tolerable but in a less accepting future may come to view as "disorders."[53]

Constraining Mothers

Courts, finally, have joined with the doctor and the state in subordinating the mother's interests to those of the viable fetus in late pregnancy. Fear of malpractice actions or even of criminal prosecutions leads doctors in these situations to recommend medical procedures that may infringe on an expectant mother's liberty or religious freedom. The record of litigation suggests that in such cases judicial power is more frequently asserted against the pregnant woman than in her favor.

The possibility of maternal–fetal conflicts came sharply to national attention in October 1986, when Pamela Monson, an indigent twenty-seven-year-old California resident, was charged with a criminal misdemeanor for her conduct during pregnancy.[54] Suffering from placenta previa, a condition that can lead to fetal death, Monson had been instructed by a physician to avoid illegal drugs, to refrain from sexual intercourse, and to seek immediate medical treatment if she began to hemorrhage. Monson allegedly failed to comply with any of these instructions and eventually gave birth to a severely brain-damaged child who died after five weeks. Several months later Monson was charged under a 1926 law that made it a crime willfully to withhold medical attendance from a child, although the charges were subsequently dropped as inappropriate under that particular legislation.[55]

Although criminal prosecutions of pregnant women are rare, disputes about a woman's liberty to make decisions adversely affecting the health of her unborn child are more common. A 1986 survey revealed that courts in eleven states had ordered obstetrical procedures to protect the fetus after the mother had refused such treatment.[56] Many of the women against whom orders were sought represented the most vulnerable sectors of society: 81 percent were black, Asian, or Hispanic; 44 percent were unmarried; and 24 percent did not speak English as their primary language. In 88 percent of the cases, it took less than six hours to obtain the court order. The survey authors concluded that complex medicolegal decisions were being made impulsively by judges who were too deferential to medical experts and without adequate legal representation for the pregnant women.[57]

The increase in such cases nationwide[58] prompted questions about the capacity of courts and the legal system to protect the civil liberties of pregnant women against physicians and state governments. George Annas has painted a grim picture of what could happen if the law fails to stand up for women's rights: "In her futuristic novel *The Handmaid's Tale*, Margaret

Atwood envisions a world in which physicians and the state combine to strip fertile women of all human rights. These women come to view themselves as 'two-legged wombs, that's all; sacred vessels, ambulatory chalices.' To some that future is now."[59] Whether or not one shares in this stark assessment, the conclusion seems unavoidable that an uncritical alliance between courts and doctors, acting under pressure of perceived medical emergencies, is not likely to produce balanced, consistent, and humane principles for resolving maternal–fetal conflicts. More promising was a statement by the American College of Obstetrics and Gynecology in August 1987 that using judicial authority to order procedures not desired by pregnant patients was a violation of the patients' autonomy.[60]

Reconstructing the Family

What makes a family? Is it genetics or a set of social characteristics not obviously linked to biology that best qualifies an adult to care for a child? The two-year custody battle over young Jessica De Boer, which pitted her biological parents in Iowa against a couple who had reared her from birth in Michigan, illustrated how open-textured these questions remained even in the 1990s, despite years of legal involvement with these issues. Other contemporaneous decisions—such as one to sever a child's relations with biological parents whom she had never known and another to remove a child from the custody of a lesbian couple—showed continuing uncertainty over what, in today's mobile and sexually liberated society, truly binds adults and children into a family.

To the continuing disputes over adoption and custody, technology has added a new dimension by permitting a conscious, premeditated separation of biological from social parenting. With a little technological help, an infant in the late twentieth century could come into the world to find no fewer than five individuals claiming some or all of the rights of parenthood: the genetic mother and father (egg and sperm donor, respectively), the social (or adoptive) mother and father, and the gestational (or surrogate) mother. The Baby M case, which first made a public issue of surrogacy, illuminated both the strengths and weaknesses of courts as places for dealing with the localized and contingent expressions of more wide-ranging transformations in reproductive behavior.

The "surrogate mother,"[61] who conceives by contract and carries the child to term only to give it up for adoption, emerged in the 1980s as an unexpected new consumer of the century-old technique of artificial insemination. By the time surrogacy arrived on the legal scene, artificial insemination by donor (AID), the practice of impregnating a woman with sperm from a donor other than her husband, had been in routine use for some two generations.[62] AID was the safe and socially accepted response to a husband's infertility or, in some cases, to a risk of producing offspring with serious genetic defects. Introducing a third party—the sperm donor—into the procreative process generated numerous disputes about the relationship of the AID child to each of its two "fathers," but the important issues were largely settled by the time courts were asked to consider the Solomonic problem of two women claiming to be "mothers" of the same child.

With occasional backslidings, courts in AID cases had generally concluded that the child's interests should be put ahead of other considerations. Some early decisions had classified AID as a form of adultery, holding that a child begotten by a party other than the mother's husband should be considered illegitimate.[63] This view, however, gave way to the more enlightened position that children born through AID are the husband's legitimate offspring as long as he has consented to the insemination.[64] Progressive judicial decisions continued to expand the rights of nonbiological fathers in these cases, even granting custody and visitation rights to the mother's ex-husband who was not genetically related to the AID child.[65] Nevertheless, biological determinism never completely disappeared from the judicial mind, as is evident from a case like *C. M. v. C. C.*[66] Here, the court granted the father's request for visitation rights, over the mother's strenuous objections, on the basis of a long-standing judicial policy "favoring the requirement that a child be provided with a father as well as a mother."[67]

Technologically, the Baby M story, which riveted the nation's attention for seven weeks in 1987, involved nothing more novel than the birth of another AID child. What rendered the case extraordinary were the changed legal and social circumstances that caused a routine medical procedure to take on the trappings of prime-time drama and morality play: for here it was a woman (not a man as in AID cases) who had allowed her genes and her body to be appropriated by an unrelated, childless couple. The story began when Mary Beth Whitehead and William Stern signed a contract under which Whitehead agreed to be artificially inseminated with sperm donated by Stern, to carry the child she so conceived to term, to surrender the child to Mr. and Mrs. Stern after birth, and to relinquish her own parental rights. In return, Whitehead, a twenty-nine-year-old married woman with two children of her own, was to receive $10,000 plus all her medical expenses. The agreement seemed at the outset one more, barely visible move in the tacit but growing social acceptance of surrogate births,[68] but the situation changed overnight when Whitehead gave birth to "Baby M" and then decided that she could not fulfill her bargain with the Sterns.

The resulting controversy presented the New Jersey courts with a choice between a biological (potentially reductionist) and a social (more holistic but also more prone to class bias) model of motherhood, much as earlier AID cases had done with respect to fatherhood. Judge Sorkow of the New Jersey Superior Court opted without reflection for the social model in his initial ruling in the case. He refused to apply state adoption laws so as to invalidate the alleged contract, because "surrogacy was not a viable procreation alternative and was unknown when the laws of adoption were passed."[69] Accepting the Stern–Whitehead contract as valid, Sorkow went on to determine Baby M's "best interests" as if in a traditional custody hearing. Not surprisingly, this calculation led him to award custody, as well as full parental rights, to the economically and educationally more advantaged Stern couple.

On appeal, the New Jersey Supreme Court read the applicable state laws in a light vastly more favorable to Whitehead.[70] The high court determined, to start with, that the surrogacy contract was invalid because it conflicted with numerous New

Jersey statutes and public policies governing adoption and parental rights. Having nullified the agreement, the court upheld the award of custody to the Sterns but restored Whitehead's maternal rights and made provisions for her to continue visiting the baby.

Biological considerations, which had been absent from Sorkow's disposition of the case, underpinned the higher court's legal interpretation. Stern had argued that invalidating the surrogacy contract would deny him equal protection by unjustly advancing Whitehead's interests at the expense of his own biological claim to parenthood. The court, however, warned against a too easy equation of Stern's position as sperm donor with Whitehead's position as genetic and gestational mother.[71] In the court's judgment, the nine months invested in a pregnancy sufficiently distinguished the woman's contribution from the man's in childbearing. Overall, the decision won praise for having withstood the class bias that had led the lower court to favor the relatively affluent Sterns and to dismiss Whitehead's claims to motherhood. Others, however, were troubled by reasoning that seemed to perpetuate the myth of pregnancy as a blissful state and the "official" image of women as mothers.[72] Full equality for women, a goal of abortion-rights litigation since *Roe,* required that Whitehead's rights be recognized as on a par with Stern's; however, vindicating these rights through an insistence on the woman's special role in reproduction and childbirth could not fail to arouse misgivings.

A later California case, *Anna J. v. Mark C.,*[73] underlined in yet starker form the risks of importing judicial understandings of biology into complex policy issues about reproduction and parenting. The conflict in this case was between the genetic mother, Crispina Calvert, and the gestational mother, Anna Johnson, who for a fee of $10,000 had carried Calvert's fertilized ovum to term. Ruling that Calvert was the mother under California law, the court of appeal held that genes were the appropriate basis for determining motherhood. George Annas, in a devastating commentary, has deconstructed the legal and biological boundary-drawing by which the court strategically picked and chose among competing bodies of knowledge and assertions of expertise to reach its result.[74] For instance, the court accepted the highly contestable claims of expert witnesses that "genes influence tastes, preferences, personality, styles, manners of speech and mannerisms."[75] At the same time, it rejected as insufficiently reasoned a recommendation of the ACOG Ethics Committee that gestation rather than genetics should define motherhood. By opting for a genetically deterministic view of human behavior and rejecting ACOG's opposing viewpoint, the court tacitly mobilized the authority of the law to support a particular vision of biological and social roles.

Law in the Networks of Biology and Society

The legal controversies discussed in this chapter develop and refine some of the themes we have previously encountered in looking at judicial responses to technological change. First, these cases again confirm the important role that courts play in the early stages of accommodating a technological innovation into its social and cultural context. Disputes come to the courts before, not after, other governmental institutions have had the opportunity or shown the inclination to consider them. Judges therefore have a particularly authoritative voice in shaping social policies for new technologies. In the ideal situation, landmark court decisions can be expected to provide the impetus for other actors to assume the obligations of policymaking. Thus, the Baby M trial greatly increased public awareness of surrogacy and advanced the issue to the head of the legislative agenda in many states at once.[76] Arkansas, for example, soon enacted a law providing that a married couple using an unmarried surrogate would be the child's legal parents without any need for adoption. With regard to *in vitro* fertilization, legislative torpor again left the courts, almost by default, as the primary custodians of the rights of embryos and their progenitors.

Second, *Roe v. Wade* and subsequent abortion cases displayed the risks and benefits of using constitutional law to cement radically changed social expectations surrounding reproductive technologies. Just as a scientific discovery—for example, the double-helix structure of DNA—cannot on its own structure a whole industry, so too an abstract constitutional principle cannot, without more, provide the infrastructure for a whole world of changed behaviors and expectations. *Roe*, for widely discussed reasons, was perhaps a particularly vulnerable armature on which to hang an authentic social revolution. At the same time, it seems clear in retrospect that constitutional law, not state-by-state legislation, was the only political vehicle through which women's sexual emancipation could have been asserted in the early 1970s. How surprising is it that another twenty years went by before important segments of the legal and political communities could join together with sufficient strength to reaffirm *Roe*'s central holding? It is worth recalling that *Brown v. Board of Education of Topeka*, a decision whose constitutional rightness has rarely been questioned, did not accomplish school desegregation overnight. *Brown v. Board*'s injunction to proceed with "deliberate speed" gave only a pale foreshadowing of the long pains of the civil rights movement, school busing, affirmative action, and the backlash against all these that were to come in the forty years following the decision. More prosaically, as argued in the preceding chapter, *Diamond v. Chakrabarty* did not immediately establish a social consensus in favor of recognizing property rights in gene-engineered animals. Looking at *Roe* as but one component in a revolution of thought about biology and sexuality, human procreation, the rights of the fetus, and women's roles in society induces greater humility in assessing the decision's significance. Writing it off as a misguided turn in constitutional jurisprudence fails to capture the decision's centrality in shoring up a new social order that was building slowly but surely through the latter half of the century.

Third, the cases on reproductive technologies call attention to the lack of reflexivity and critical acumen in judicial thinking about biology as an explanatory resource in legal decisionmaking. Although almost all the litigation in this area deeply entrenches on the boundary between the natural and the social worlds, and could in principle provoke profound deliberation on human nature, agency, and control, decisions such as *Baby M* and *Anna J.* are remarkable for their unexamined, taken-for-granted assumptions about what is biological and what is social. That the law actually plays a part in constructing the public understanding of these domains (for example, by elevating the "biology" of genetics above that of gestation in *Anna J.*) is an insight absent from most judicial reasoning. Only briefly in *Casey* did a court glimpse the possibility that human beings, as members of both the natural and social worlds, can actively engage in creating and recreating the boundaries that define and constrain their self-images and actions.

9

Definitions of Life and Death

For seven years, thirty-two-year-old Nancy Cruzan lay comatose in a Missouri hospital, her condition described by doctors as a "persistent vegetative state" and her "life" artificially sustained through feeding tubes. Then, on June 25, 1990, her plight was reviewed by the nation's highest court. In *Cruzan v. Director, Missouri Department of Health,*[1] the Supreme Court ruled that the State of Missouri could lawfully deny a request by Cruzan's parents to disconnect the tubes, thus allowing their daughter to die. The reason: Missouri law requires "clear and convincing" proof that a comatose patient would have wished to discontinue life-sustaining treatment, and Cruzan, while still conscious, had failed to express her wishes in a manner that satisfied this legal standard. Although the decision prolonged the Cruzan family's distress along with Nancy's life,[2] it was hailed by many as a monumental achievement in constitutional law. The Court had recognized for the first time that personal liberty includes the right not to perpetuate a cognitively empty life.

But how did such a traditionally private issue as the treatment rights of dying patients find its way from the secluded medical arena to the glare of constitutional litigation, and how did it come to be framed as a question of individual liberty? More generally, how effectively did the legal process deal with the nested questions of ethics, economics, and social policy surrounding the use of life-sustaining technologies? This chapter tells two stories about the evolution of right-to-die cases, one involving adult patients and the other severely handicapped newborns. In the former, courts appeared at their institutional best, gradually unpacking issues of individual rights and social responsibilities and providing analytic guidance to other policymaking bodies. In the latter, legal deconstruction proved less effective, showing again the limits on the capacity of the courts to challenge socially dominant perceptions of human agency and the power of technology.

The legal issues surrounding the right to die were first defined in contests between physicians who feared liability and patients or their families who viewed heroic treatment as futile. Over time, questions of immediate professional concern to physicians, such as the appropriate criteria for disconnecting a respirator from a comatose patient, became the springboard for a broader judicial inquiry into what kind of life is worth living, who should speak for patients who cannot speak for themselves, and when treatment becomes a restraint on liberty. With the aging of the American population and the escalating cost of health care, the legal dimensions of these questions imperceptibly shaded into larger economic and public policy considerations. To what extent can even the richest society afford to prolong the lives of terminally ill citizens? Should seriously ill newborns be kept alive irrespective of cost and parental judgment? Should right-to-die decisions take account of the potential for organ harvesting and transplantation? And can a space be carved out for medically assisted suicide without compromising the commitment of physicians and the state to preserving life?

At first debated by lawyers and ethicists, these questions eventually moved out of the judicial and philosophical domains to engage the attention of legislatures, health care providers, and governmental policymakers. Courts in the process relinquished their temporary monopoly on legal rules concerning life-prolonging technologies, and judicial interpretations of the right to die became enmeshed in a complex tapestry of social adjustments to the newly public, technologically assisted, rituals of dying.

From Deathbed to Courtroom

Changes in the way people enter and leave their lives constitute a significant chapter in the social history of the twentieth century.[3] Since 1900, the process of dying has moved relentlessly away from the private care of homes and families into the public stewardship of hospitals and institutions for the aged.[4] Death steals upon its victims later, less openly, and often after long, degenerative illnesses. With dramatic advances in surgery, organ transplantation, drug therapy, and resuscitation, there is scarcely any disease that does not lend itself to some form of interventionist medical endgame. Chance and nature play an ever less controlling role as human beings approach the inevitable moment of death.

For centuries, the Judaeo-Christian tradition made amends for the impermanence of the body by celebrating the immortality of the soul. Western art, music, and literature triumphantly proclaimed that, although the physical frame may die, the believing spirit, escaping earthly bonds, will enjoy everlasting life. In our own time, however, technology has ironically reversed the classical presumption. Attached to the artificial respirator and feeding tube, the body remains physically functional long after the end of meaningful cognitive existence. By introducing sudden and still unresolved ambiguity into the age-old meaning of "death,"[5] these technological devices challenged not only social but legal norms. Litigation revealed the need for new understandings about whose policies and principles should govern treatment decisions at the end of life—the sick and the dying, their families and physicians, or impersonal institutions, such as hospitals, courts, and legislatures. The issues quickly transcended scientific fact-finding and medical judgment, for neither biologists nor doctors could decide on their own at what point a patient ceases "to be an individual in need of medical care and become[s] a corpse."[6] Physicians who dared to think otherwise sometimes unwittingly exposed themselves to the threat of civil and criminal liability, although, as the medical ethicist Norman Fost notes, no doctor has "ever been found liable for deliberately withholding or withdrawing any life-sustaining treatment from any patient for any reason."[7] Not surprisingly, the medical community turned to the law for a clearer articulation of patients' rights and a redrawing of the boundary between responsible and irresponsible medical practice.

This instrumental use of law, however, entailed significant institutional consequences, as a wide range of formerly invisible medical decisions were converted into public legal disputes, straining judicial resources and placing courts in the awkward position of second-guessing physicians. The legal system was forced under the onslaught of new cases to remedicalize decisions about death and dying. The challenge was to develop rules of conduct that would respect the rights of patients, as revealed and refined by courts, and yet would permit physicians and hospitals to resume their responsibilities with a clarified understanding of the boundary between life and death, and without fear of legal reprisal. The state's role in protecting life also had to be reevaluated, especially when it conflicted with shifts in the rights of dying patients. By the mid-1980s, the psychological as well as economic impacts of modernizing death grew too substantial for case-by-case resolution, and it became necessary for both law and medicine to consider new approaches to alleviating the burdens of dying.

Framing the Problem

Just as *In re Baby M* signaled the need to reconfigure the meaning of "motherhood," so the *Quinlan*[8] case in New Jersey and the nearly contemporaneous *Saikewicz*[9] decision in Massachusetts brought the emerging legal and ethical issues surrounding life-sustaining technologies to the forefront of public awareness.[10] Converting an incremental, almost invisible, process of social change into a recognized public problem,[11] these cases rightly came to be regarded as milestones in the law concerning treatment decisions for adult "silent patients."[12] Together they mapped out a series of questions concerning life-prolonging technologies, and their supposed beneficiaries, that were to occupy courts and legislatures for many years to come.

As happens so often in the law, a small though tragic event

precipitated unforeseeable consequences. On the evening of April 15, 1975, twenty-one-year-old Karen Ann Quinlan set out in apparently perfect health and spirits to attend a party with friends. Later that evening, for reasons probably related to an adverse drug reaction, she lost consciousness and stopped breathing for at least two fifteen-minute periods. She was admitted to a New Jersey hospital where, after a number of emergency measures failed to revive her, she was placed on a respirator and given nourishment through a feeding tube. Thereafter she remained in a deep coma later diagnosed by experts as a "chronic persistent vegetative state" without any cognitive function. Quinlan, however, was not clinically "brain dead"; she retained some brain stem function and responded minimally to external stimuli such as light and sound.

Most of Quinlan's doctors agreed that her chances of returning to a fully sapient existence were exceedingly remote, but they also expressed a professional judgment that she should be maintained on respirator support, since her condition did not meet the standard of brain death.[13] Her father, Joseph Quinlan, a deeply religious Catholic, contested this decision and, supported by family members and the church, petitioned the courts to recognize his right as his daughter's guardian to demand the respirator's withdrawal. In a decision that became the touchstone for later right-to-die cases,[14] the New Jersey Supreme Court concluded, first, that Quinlan had a constitutional right to privacy that allowed her to make decisions about the treatment of her own body. Her father, the court further held, was entitled as a properly qualified guardian to assert this right on Quinlan's behalf. Third, the court recommended that in the future such decisions should be made by pluralistic "ethics committees" composed of physicians, social workers, attorneys, and theologians. These committees would "screen out" any possibly unworthy motives of families and physicians and help to resolve the majority of cases without need of judicial intervention. If such a procedure were in place, the court noted, judicial review would come to be regarded as inappropriate, "not only because that would be a gratuitous encroachment upon the medical profession's field of competence, but because it would be impossibly cumbersome."[15]

Following this decision, Quinlan's doctors agreed to withdraw their patient from the respirator, apparently expecting that she would soon die. But Quinlan confounded most medical predictions, as well as the law's construction of her physical state as near death, by surviving in a comatose condition for the next ten years.[16] Her home during this period was "a locked, dimly-lit, 8-by-10-foot nursing-home room with the shades drawn,"[17] where she was frequently visited by her parents but probably did not receive extraordinarily attentive nursing care. She died on June 11, 1985, without any sign of regaining consciousness.

At almost the same time as *Quinlan* was being deliberated in New Jersey, the Massachusetts Supreme Judicial Court was asked how to make life-and-death decisions on behalf of patients who, unlike Quinlan, had never fully participated in the networks of normal human interaction and communication. The issue was whether chemotherapy could lawfully be withheld from Joseph Saikewicz, a mentally incompetent patient at a state hospital who was suffering from leukemia. Retarded since birth, Saikewicz was sixty-seven years old at the time of diagnosis but had the I.Q. of a two-year-old. Doctors testified that he would not be able to understand the purpose of the treatment or its side effects, which could be painful and even fatal. If denied chemotherapy, Saikewicz was expected to die in a matter of weeks or months without experiencing special discomfort.

The *Saikewicz* court agreed with *Quinlan* that ethics committees could play a valuable role in advising probate judges on decisions about prolonging the life of incompetent patients. But the Massachusetts court sharply contested the New Jersey high court's suggestion that consultation with and deference to the medical profession could take the place of judicial oversight. Far from dismissing judicial review as "a 'gratuitous encroachment' on the domain of medical expertise," the *Saikewicz* court saw right-to-die cases as peculiarly the preserve of the judicial branch, which alone could provide the "detached but passionate investigation and decision" that these disputes demand.[18] This seemingly imperialistic stance reflect-

ed not so much a desire to override medical opinion as a determination to control the legal standard for making decisions.
The standard that the Massachusetts court selected was that
of "substituted judgment," which required the surrogate
decisionmaker to make the decision "which would be made by
the incompetent person, if that person were competent, but
taking into account the present and future incompetency of
the individual."[19]

The Supreme Judicial Court hoped by this means to give
controlling weight to the patient's viewpoint, thereby avoiding
the pitfall of paternalistic decisions by medical experts. In
practice, however, the substituted-judgment standard proved
almost impossible to implement, since no fully competent
decisionmaker could hope to enter so completely into the
thought processes of an incompetent person as this standard
demanded.[20] *In re Storar,* a case closely resembling *Saikewicz,*
afforded the New York Court of Appeals an opportunity to re-
ject the substituted judgment test (the court compared it to
asking whether "if it snowed all summer it would then be winter") and opted instead for the "best interests" standard normally applied to decisions involving children.[21] Despite greater
judicial familiarity with this approach, it too had its problems,
as we shall see, for it failed to recognize that medical and judicial biases in favor of treatment could overwhelm dispassionate appraisals of the situation from the patient's standpoint.

Constructing the Patient

In both *Quinlan* and *Saikewicz,* the courts took up the complex
process of constructing the patient that begins in the medical
arena with the earliest acts of diagnosis and treatment. These
opinions represented the judicial system's first tentative steps
toward determining the moral status of human beings hover-
ing on the borderline of life and death. Clarifying this status
took no small expenditure of judicial energy. The courts had to
find legally meaningful ways of describing the incompletely understood dependencies between patients and physicians, between patients and their life-support systems, and between

patients and the state. Each step in this complex process of social construction involved false starts and wrong turns that had to be modified or undone by later decisions.

Doctors and Patients

The prime rationale for judicial intervention in *Saikewicz* and other right-to-die disputes was the supposedly unique qualification of courts to weigh conflicting values and protect the rights of the weak and powerless. The conflicts that came to court involved obvious inequalities. On one side were patients in varying degrees of physical and mental debility, whose despairing claims (sometimes asserted by surrogates) were based on the values of autonomy, privacy, and human dignity. On the other were physicians and medical institutions who claimed superior authority by virtue of technical know-how and a collective, state-sanctioned commitment to saving lives. Under these circumstances, the institutional duty of courts was to safeguard the patient's rights and interests—in short, to act as a counterweight to more dominant actors, including family members, whose wishes did not invariably coincide with the patient's.

In its report on forgoing treatment, the President's Commission for the Study of Ethical Problems in Medicine cast doubt on the capacity of courts to perform this function effectively in light of their well-known hesitancy to second-guess medical experts: "the question typically addressed is whether the particular treatment chosen is the right one. Since an answer to this question would normally require substantial understanding of the patient's evolving medical condition, which the courts lack, they may simply defer to the recommendation of the treating physicians."[22] The leading right-to-die cases confirmed these suspicions, demonstrating the apparent inability of judges to mount a fully independent and critical review of testimony by treating physicians. The court's constructions of the patient's mental and moral status unquestioningly incorporated physicians' technical judgments—and therewith, of course, the medical community's embedded normative presuppositions about the patient's condition and competence.

The *Quinlan* decision, for instance, turned on the court's assessment that the patient was irreversibly comatose and wholly dependent on a machine in order to continue breathing; her treatment was therefore conceptualized as an artificial prolongation of dying, a technological intervention holding no promise of ultimate recovery. In Robert Burt's view, these conclusions were based on a selective and possibly unfounded reading of the medical testimony. Both the trial judge and the state supreme court disregarded the testimony of the only physician who had actually attempted to wean the comatose patient from the respirator and who had concluded, correctly in the event, that she might survive even without mechanical assistance.[23] The high court was also on questionable ground in analogizing Quinlan's condition to that of "a competent patient, terminally ill, riddled by cancer and suffering great pain."[24] Although this characterization strengthened the impression of the patient's utter hopelessness, the medical testimony before the court suggested that Quinlan was not in fact capable of feeling pain.

In language drawn from the sociology of science and technology, the court here chose to disregard the "interpretive flexibility"[25] disclosed by the medical testimony in order to simplify its characterization of the patient. "Prolongation of death" denied agency to Quinlan (who was now seen as having departed from life) and hence had very different legal and medical connotations from "treatment of disease." One wonders whether the *Quinlan* court would have been as ready to discontinue treatment if it had focused on the evidence that the patient was *not* in pain and that her condition was *not* terminal.[26] By contrast, in the *Storar* case mentioned above, adoption of a treatment model led the New York Court of Appeals to agree with the patient's physicians that continued blood transfusions would be in his best interests, although his mother opposed the treatment, and the evidence showed that he had only a few months to live and found the procedure extremely disagreeable.[27]

In *Saikewicz*, as well, judicial opinion was molded to an untoward degree by the preferences of medical experts, but without the editing of interpretive flexibility that took place in

Quinlan. Saikewicz' doctor initially deferred to the judge, saying that he himself did not have the "deep knowledge" required to make the treatment decision.[28] As soon as the judge expressed his inclination to continue chemotherapy, however, both the doctor and Saikewicz' court-appointed guardian rushed to inform him that the patient would become unmanageable and that the procedure might actually shorten his life. Their onslaught caused the judge to waver and then change his mind. The final order stated: "After a full hearing with medical specialists and doctors being present and their testimony being taken, the Court determines and adjudges that chemotherapy treatment should not be given at this time."[29] The deliberative tone of the order masked a conclusion that the hearing record makes plain: the judicial decision about Saikewicz' fate was in reality hardly more than a legal rubber stamp for the wishes of the patient's medical caretakers.

Problems of characterization proved no less difficult for courts when fully conscious, intelligent patients were subjected to extraordinary medical measures. The case of Elizabeth Bouvia was among the most poignant of the many similar conflicts that arose around the country during the 1980s. A study in trial and error, the California judicial system's handling of her complaints was condemned by one observer as even more insensitive than her treatment at the hands of the medical profession.[30]

Bouvia's problem was that she was mentally capable but physically incapacitated. Completely paralyzed by cerebral palsy, she could not take care of any bodily functions without assistance. In 1983, separated from her husband and living with her family, the twenty-six-year-old woman decided that she could not face the rest of life as a helpless invalid. She had herself admitted to a California hospital where she announced her intention to fast to death. The hospital, however, refused to participate in an apparent suicide attempt and insisted on force-feeding Bouvia through a nasogastric tube. When she sought a court order to end the feeding, a California superior court upheld the hospital.[31] Three years later, following another bout of hospitalization and force-feeding, a state appeals court finally honored her wish to have the feeding tube disconnected. On this occasion, Bouvia had to assure the hospital she did not plan to refuse nourishment.

Bouvia's existence remained sadly constricted even after she won the right to control her treatment. Ten years after her 1983 legal victory, she was living a life of nearly complete isolation in an $800-a-day hospital room in California. An attempt to starve herself to death in 1987 ended when she learned that it could take several weeks to die and she could not endure the pain and other side effects. In a rare interview in 1993, she told a journalist that she still wished to "die peacefully" but that, failing an early death, she was "trying to work on getting out of the hospital."[32]

State courts began in time to view the rights of competent but disabled patients in a more sympathetic light. A California court held in 1984 that an incurably ill adult patient could refuse life-supporting treatment although his condition had not been diagnosed as "terminal."[33] In a subsequent New York case, a mentally alert eighty-four-year-old patient in a nursing home was granted the right to starve herself to death.[34] A Colorado court not only permitted a man of thirty-four, paralyzed from a drug overdose, to stop taking food and water, but also ruled that he could not be moved from the hospital where he had been receiving treatment.[35] The judge set aside the hospital's concern about participating in suicide on the ground that death in this case would be a "natural consequence" of the patient's illness. *Cruzan* completed this movement toward greater respect for the patient's autonomy. The Supreme Court held that the liberty interests protected by the due process clause of the U.S. Constitution included the freedom to refuse unwanted medical treatment. Notably absent from this reasoning was the concept of privacy invoked in *Quinlan* and in many later state right-to-die cases.

While illustrating a libertarian trend, these decisions also marked the tragic limits of judicial power to address the patient's true needs and intentions. Courts could facilitate dying, with appropriate legal constructions of the patient's state, but they could not restore to life the meaning and freedom so desperately craved by intelligent, sentient, and disabled human beings. Judges granted pleas for removal of

artificial life supports in cases in which death was not necessarily the patient's driving motive. Larry McAfee, a quadriple-
gic former mechanical engineer in Georgia, dramatized this
point when he chose not to exercise his hard-won legal right to
turn off his respirator.[36] The publicity generated by his battles
with the courts produced enough social support for McAfee to
decide that he would continue at least temporarily his excruci-
atingly circumscribed existence.

Letting Machines Rule

Following *Quinlan,* right-to-die litigation entered upon a generally unproductive detour as state courts tried to find principled
reasons for determining whether the removal of a feeding tube
should be viewed as legally different from detaching a respirator or discontinuing other forms of treatment. In one or two
early cases, artificial nutrition was mandated on the ground
that courts should never order cessation of a procedure that
was guaranteed to terminate the patient's life. These decisions
tracked the belief of some medical ethicists that denying food
and fluids to patients, even comatose patients, was inconsistent with fundamental moral values and the ethical practice of
medicine.[37]

Arguments that withholding nourishment was more "final"
than ceasing other treatment methods ultimately failed to persuade the legal community (we may compare here the similar
difficulties, as seen in Chapter 8, of "privacy" as a rhetorical
construct for accommodating abortion decisions). George
Annas, for example, argued that withdrawal of ventilation, di-
alysis, or chemotherapy from a dying patient was no less final
in most instances than withdrawal of food.[38] Moreover, the
focus on specific technologies seemed to distract judicial attention from keeping the patient's needs and social context at
the heart of legal analysis. Thus, in *Brophy v. New England
Sinai Hospital, Inc.,* a Massachusetts probate court showed
more concern with analyzing the physical properties of a gas-
trostomy tube than in asking about the patient's wishes. Having determined that the tube was "non-intrusive" (a judgment
made possible only by separating the tube from the entire cir-

cumstances of its use), the judge ordered feeding and hydration to continue, even though it was shown that Brophy, if competent, probably would have chosen to forgo treatment. The court's action not only went against the wishes of the patient and his family but entailed massive expenses. According to one contemporaneous account, the cost of keeping patients like Brophy physically alive could run as high as $11,000 a month.[39] The Massachusetts Supreme Judicial Court eventually reversed the lower court's decision,[40] adding to a small but growing number of cases in which the use of feeding tubes was found to be no different from other forms of life-prolonging treatment.

Legal boundary-drawing over the characteristics of treatment methods also exposed cultural differences in the professional languages of law and medicine. Formulations carrying one meaning for judges could carry quite another for physicians. Just as biologists and lawyers disagreed about the meaning of "major environmental impact" in relation to genetically engineered bacteria (see Chapter 7), so doctors and judges differed about what constitutes "heroic" or "extraordinary" treatment. The President's Commission noted that such terms tend to have an "unfortunate array of alternative meanings":[41] what a doctor sees as "standard procedure" may well be branded "extraordinary" by a judge, leading to a wholly different appraisal of the need for the technique. The Commission therefore recommended abandoning the extraordinary/ordinary distinction as a basis for legal decisions and replacing it with an inquiry into the relative benefits and burdens of different treatments from the standpoint of the patient.[42] The proposal marked a salutary refocusing of attention on the central figure in the drama of dying, but it could not cure the underlying problem of divergences between the specialist's and the lay person's views of the need for technology or the moral status of a technologically extended life.

The Supreme Court of New Jersey was the first state high court to permit discontinuance of feeding in a case involving an eighty-four-year-old woman residing in a nursing home, whose condition was described as "severely demented."[43] Although the ruling was restricted to elderly nursing home patients, the finding that there is no legal difference between artificial feeding and other forms of life support became generalizable to other classes of patients. It remained for the U.S. Supreme Court to decide definitively in *Cruzan* that refusing forced nutrition and hydration was within the individual's constitutionally protected sphere of medical self-determination. Justice O'Connor stated the point most explicitly in her concurring opinion: "Artificial feeding cannot readily be distinguished from other forms of medical treatment."[44] At the same time, the Court felt that the finality of such decisions justifies state laws which, like Missouri's, demand a high level of proof from surrogate decisionmakers who seek to discontinue nutrition and hydration on behalf of a silent patient.

Even as judicial doctrine was moving in the direction of *Cruzan,* some state legislatures seemed intent on restoring a sharp distinction between terminating nourishment and other forms of life support in their right-to-die statutes. In these jurisdictions the individual rights-based approach of the courts seemed to be on a collision course with community sensibilities concerning the special virtues of feeding the dying. A similar assertion of community values accounts for the legal and medical impasse over the treatment of defective newborns.

The Patient as Citizen

While courts were constructing the patient in relation to the human and mechanical paraphernalia of treatment, other state actors, both legislative and executive, approached the right to die from the standpoint of the state's responsibility to save lives and promote the welfare of citizens. Legislative efforts provided the instruments for removing the great majority of right-to-die cases from the courts. Executive intervention on behalf of seriously ill newborns, by contrast, pointed ahead to one of the unresolved dilemmas of the 1990s.

Empowering Patients: The Role of Legislation

State legislatures displayed a broad-based commitment to giving formal recognition to the wishes of dying patients and their families. Many states provided legal authority for family members or others to make medical decisions on behalf of sick or incapacitated adults, thereby removing the need to determine who should speak with the patient's voice in each case.[45] To be sure, the appointment of an appropriate surrogate did not in itself resolve the problem of ascertaining the patient's wishes. The Supreme Court's holding in *Cruzan* heightened public awareness of the utility of "advance directives"—legal documents that would permit healthy adults to spell out how they would like to be treated if they ever lost the capacity to make medical decisions for themselves. State statutes became the primary vehicle for enforcing these directives.

In the ten years following *Quinlan,* a majority of states enacted so-called natural death acts to provide individuals a recognized legal means of refusing life-sustaining treatment. Ten states adopted such laws from 1976 through 1980, an additional thirteen from 1981 through 1984, and thirteen more in 1985 alone.[46] The pace of enactment said much about the growing social acceptance of the principles of self-determination in dying, but the laws did not always successfully resolve the problems raised by the early right-to-die cases, and they created their own problems of interpretation.

Most natural death acts include certain basic provisions: recognition of an adult individual's advance directive concerning medical care; legal immunity for medical caregivers; a form for the declaration, often providing for individual modifications; and definitions of relevant terms.[47] The commonest form of advance directive is the "living will," a written declaration of a conscious and competent adult's wish to forgo treatment in the event of total incapacitation by disease or accident. Developed in 1930 by Luis Kutner, a crusading civil rights lawyer, the concept of the living will took two decades to gain social acceptance and even longer to achieve legal recognition.[48] Nonetheless, the documents proved immensely popular and were distributed to thousands, even millions, of interested citizens by the early 1980s.[49] Some months after *Cruzan,* a right-to-die advocate estimated that the decision had prompted hundreds of thousands more to sign such documents.[50]

For all their popularity, living wills were surrounded by ethical and legal uncertainties. To begin with, no directive drawn up in advance could incorporate the individual's actual contemporaneous response to physical or mental incapacity. In respecting such a document, the medical caregiver is at best giving effect to a projection made by the individual on the basis of incomplete and hypothetical knowledge—not to an authentic act of self-determination. The "durable power of attorney" avoids some of these weaknesses. This instrument is used by one person (the "principal") to confer on another (the "agent") legally recognized authority to make decisions on the former's behalf. The power so conferred is "durable" because it continues even after the principal is incapacitated. By specifically addressing the issue of medical treatment, a durable power of attorney can provide invaluable evidence of the principal's intentions. At the same time, it is more flexible than a living will because it permits the lawful agent to consider not only the principal's previously expressed wishes but also the actual medical circumstances at the time of decision. Because of these advantages, lawyers and policy analysts recommend wider use of durable powers of attorney in medical decisionmaking,[51] but the status of this instrument in right-to-die cases had to be clearly established in state legislation, and by the late 1980s only a minority of states had taken such action.[52]

Natural death acts may contain significant limitations on the right to die. Some provide, for instance, that artificial feeding and hydration may not be withheld from terminally ill patients. Since artificial feeding emerged as "the major right-to-die issue of the '80s,"[53] such provisions greatly restrict the law's range of application. Most natural death acts also contain a clause invalidating declarations made by pregnant patients, although this provision is sometimes limited to cases in which the fetus is viable.

These statutes may also be confusing or unnecessarily restrictive in limiting the procedure and form for making advance declarations.[54] The majority of state laws, for example, permit withdrawal of treatment only when the patient is certifiably in a "terminal condition," and California's Natural Death Act of 1976 further provides that an advance directive is valid only if it is re-executed after the patient is diagnosed as being in such a condition.[55] Because of this provision, William Bartling, a patient suffering from lung cancer and emphysema and maintained on a respirator, found himself trapped in a seemingly hopeless impasse between medicine and the law. Bartling's physicians refused to declare him terminally ill, so that he was not authorized to execute a binding directive under the natural death act. His hospital meanwhile refused to comply with his treatment requests on the ground that he was depressed and incapable of making enforceable decisions. The court of appeal determined in the end that Bartling was indeed competent to refuse treatment and that he could do so without invoking the natural death act.[56]

Defective Newborns and the Grip of "Treatment"

Caring for seriously ill newborns was widely recognized as a growing problem in the biomedical community by the early 1980s. In its report on forgoing treatment, for example, the President's Commission recommended that such decisions be made more explicitly, in consultation with an intra-institutional review panel such as an ethics committee, and that therapies be classified as "clearly beneficial," "ambiguous," or "clearly futile." The Commission also concluded that the parents' treatment preferences should govern in all cases except when physicians concluded that a therapy was clearly beneficial but the parents refused treatment.[57] The assumption evidently was that physicians would recommend treatment more often than families.

In 1982 these relatively closed professional deliberations burst into public view when a newborn boy with a blocked esophagus and Down's syndrome was allowed to die at the parents' request, and with court approval, in an Indiana hospital. Responding to a public outcry, President Reagan requested the attorney general and the Department of Health and Human Services (DHHS) to enforce federal laws protecting the handicapped in future cases of this kind. Baby Jane Doe provided the needed spur to action. She was born in a New York hospital on October 11, 1983, with severe multiple birth defects, including spina bifida, hydrocephalus, and micro-cephaly.[58] Medically, legally, and ethically, hers was the pro-verbial "hard case." Her doctors agreed that without surgery to close the lesion in her spine she would die of infections within a very short time. They disagreed, however, as to how com-pletely she might recover physically or mentally following cor-rective surgery. The baby's parents refused surgery in accor-dance with the advice of a neurologist who painted a negative picture of the infant's prospects, although another doctor pre-dicted that the baby would develop some awareness of her family and surroundings.

First brought to court by A. Lawrence Washburn, a Vermont lawyer active in the right-to-life movement, the Baby Doe case triggered some two years of legal activity eventually implicating state and federal judges, Congress, and DHHS. The New York Court of Appeals denied Washburn's suit, holding that, as an outsider to the proceedings, he lacked standing to intervene in a decision made by the baby's family and qualified medical ex-perts.[59] DHHS in 1982 and 1983 issued regulations that gave the government broad powers to enter and inspect hospitals on the basis of reports that "handicapped infants" were being denied proper care. Fear that heavy-handed federal regulators would descend on hospitals in the form of "flying Baby Doe squads" and stamp out medical discretion led to further litiga-tion. After the Second Circuit Court of Appeals invalidated these regulations,[60] Congress responded in 1984 with new leg-islation aimed explicitly at limiting physician decisions to with-hold treatment from disabled newborns.[61] The following year, DHHS finally issued regulations requiring states to establish programs for the purpose of responding to reports of medical neglect.

In Fost's view, the 1985 Baby Doe regulations eliminated vir-tually overnight the problem of undertreatment for infants, mostly victims of Down's syndrome and spina bifida, who could in principle have led long and meaningful lives. Instead, the balance tilted toward overtreatment, that is, inappropri-ately intrusive care for infants with no prospects for any mean-ingful life. In an optimistic review of the achievements of hos-pital review committees, Fost concluded that group deliberations conducted by such bodies protected to some degree against overtreatment.[62] Yet cases coming before the courts in the 1990s suggest that this may have been a prematurely positive assessment. Neither courts nor medical institutions have been able to prevent continued treatment even for hopelessly ill newborns when the parents demand it and there is some institution somewhere that is prepared to offer it.[63]

Although journalistic accounts have treated these cases as involving mainly ethical questions, they take shape around a conception of "treatment" that is in fact a much more complex social construct. The word "treatment" compels action because it encompasses and gives expression to disparate yet potent values: society's desire to accord transcendental value to human life (itself a social construct in right-to-die cases),[64] medicine's commitment to the image of the heroic caregiver, parental hopes for the survival of a child, the culture's technological optimism—all these social commitments intersect in a given case with more local and contingent features of politics, such as the strength of the antiabortion movement, the interests of particular medical professionals, and the demands of individual families. I have argued in general that the strength of litigation as a policy instrument lies precisely in its capacity over time to deconstruct such black-boxed concepts as "treatment," making their components transparent in ways that permit renegotiation to suit newly recognized needs and interests. It remains to be seen whether courts will live up to this challenge in litigation over hopelessly ill newborns.

Reconsidering the Judicial Role

The law and policy concerning dying patients reaffirm the pivotal role played by the courts in the management of technological change through repeated calibrations of society's responses to new artifacts and new capabilities. The courts first elevated the issue of life-sustaining treatment to the status of a public problem. Pressed into action over a decade and a half, they developed an analytic framework for deciding who should control the process of dying, particularly in cases involving adult patients. The pathbreaking opinions of the 1970s clarified important boundaries between the physician's authority to prescribe treatment and the patient's power to refuse, as well as between the state's interest in preserving life and the individual's right to die. *Cruzan*'s ratification of the dying patient's liberty interests was in principle an important victory for the relatively powerless private individual against the institutionalized interests of medicine and the state.

Judicial performance in the years following *Quinlan* evinced some of the typical drawbacks of decentralized, court-centered decisionmaking. Judges wavered in their efforts to formulate clear guidelines for the treatment of the dying. The patient's concerns were not always carefully disentangled from the surrounding context and given proper weight. Courts either unconsciously misinterpreted expert testimony or deferred too easily to medical advice and the perceived power of machines in order to construct suitably passive patients whose interests could be protected on grounds of "privacy." Conflicting and confusing rules emerged from different jurisdictions. Even the highly acclaimed *Quinlan* decision went astray in suggesting that ethics committees should function as prognosis review boards, a role that would in no way be distinctively ethical.

On balance, however, the courts dealt with the social impacts of life-sustaining technologies in ways that furthered the development of collective norms and of new institutional arrangements with respect to death and dying. Following *Quinlan*, courts successfully delegated to ethics committees the responsibility for micromanaging right-to-die decisions. The President's Commission discovered that fewer than 1 percent of all hospitals nationwide had such committees in place, but the numbers are now substantially higher.[65] Although the impact of these bodies requires further study, there is at least anecdotal evidence that they adequately serve the functions of context-sensitive deliberation and communal norm-building that juries also provide, though often at greater expense.[66]

State legislatures, too, benefited from the prior articulation of principles by the judiciary. The proliferation of natural death acts, which formalized many of the concepts and principles set forth in the early right-to-die rulings, represented in the long run a more permanent policy solution. With state legislation generally in place, there was no longer as much need for courts to tinker with the frontiers of the common law in order to give individuals greater control over the human and nonhuman machinery of medicine. Judicial decisionmaking could concentrate instead on resolving ambiguities in relevant state enactments, a task for which courts were institutionally better adapted. Initiatives to develop new law could be focused on those areas that were excluded from coverage or left ambiguous by state right-to-die laws, for example, artificial feeding, withdrawal of life support from pregnant patients, treatment of seriously ill newborns, and possibly euthanasia.

10

Toward a More Reflective Alliance

Is the relationship between science, technology, and the law an essential alliance or a reluctant embrace, a collaboration or an unhappy marriage? Is it a "culture clash" that can be bridged only by individuals or institutions with multicultural expertise? Are scientists who participate in the legal process tolerated meddlers or essential contributors? The neat epithets of scholars and journalists scarcely do justice to the complex archaeology of modern courtroom conflicts over science and technology and the social yearnings and unrest that underlie them. Ranging from the narrowly technical to the morally divisive, from the structural and persistent to the contingent and evanescent, these controversies defy simple categorization because they embrace the totality of contemporary society's attempts to understand and control perceived threats to its stability and identity. Science and scientists are drawn into the courtroom not merely as adjuncts to legal fact-finding, but because human technological ingenuity continually gives birth to new and unruly forms of life, outstripping the equally human craving for predictability and repose.

Our task in this concluding chapter is to tease out, if possible, from the welter of law–science controversies some enduring patterns of success and failure. How does litigation advance and how does it hinder policymaking for science and technology in a democratic society? What are the most problematic consequences of using adjudication to resolve disputes among technical experts? What institutional features in law, science, technology, or the political process account for these problems, and how can or should they be remedied? Only by systematically working through these questions can we hope to assess whether fundamental changes are needed in the legal system's methods of dealing with science, technology, and social change.

Across a wide spectrum of legal controversies, there is evidence to bear out and even amplify on the complaints rehearsed in the opening chapters of this book. Courts often do appear uncertain about or resistant to quantitative methods and principles that scientists take for granted, such as concepts of causation, probability, and statistical significance. They seem even less systematically aware of modern science as a social institution, whose claims and credibility are produced through complicated negotiations within the community and with external institutions. Case-by-case adjudication leads to incoherent results in the evaluation of technical evidence, producing uncertainty for businesses and professional communities as well as for innocent victims. Judicial review provided inconsistent guidance in interpreting complex environmental statutes and exposed regulatory agencies to new crises of credibility. Tort law may well have exerted a negative impact on technological innovation and risk-taking, although corporate behavior is driven in these respects more often by perceptions of the liability system than by its actual performance.

In spite of these deficits, we have seen that courts in general and state courts in particular remain an indispensable and often appealing forum for resolving technical controversies. The cases discussed throughout this book document many instances in which courts almost by default were required to take the lead in constructing new social and political orderings around science and technology. Thus, Congress' deliberate unwillingness to act left the courts in charge of devising complex litigation procedures and rules of liability to meet the needs of victims of toxic exposure. Changes in biomedical technology and practice likewise elicited only slow and faltering responses from state and national legislatures. From surrogate motherhood to the right to die, the task of articulating policy fell in the first instance to the judiciary. Legislatures frequently held back from acting until landmark cases such as *In re Quinlan* or *Baby M* spotlighted problems whose urgency could no longer be denied. A court decision, *Diamond v. Chakrabarty*, played a seminal role in the commercialization of biotechnology. The most powerful new forensic technique of recent decades—DNA fingerprinting—began to be debated in national scientific and law-enforcement circles only after state court proceedings exposed its eminently contestable claims to reliability.

Lawsuits, we must conclude, are an essential part of the process by which American society comes to grips with the moral, material, and institutional dimensions of technological change. Doing without the courts is simply not a viable option for science and technology in the United States; litigation is too pervasive a feature of a political culture that prizes both the fine-tuning of adjudication and its quick responsiveness to individual complaints. The pressing question for policy, then, is how litigation can be made to work better when confronted by problems with significant scientific or technical dimensions. To focus this inquiry more sharply, let us first review the most salient findings from previous chapters about the legal system's performance in matters of science and technology. With these as points of departure, we will ask how courts can most productively fulfill their multiple obligations as interpreters, regulators, consumers, and, to some degree, coproducers of science and technology.

The Myths of "Mainstream Science"

Courtroom battles of experts and the perception that litigation favors the wrong side in these contests account for the unexamined though widely held conviction that "mainstream science" could dispel most of the legal system's problems in handling sociotechnical conflicts. This conviction, I have argued, rests upon fundamental misconceptions about the links between scientific and legal decisionmaking. Contrary to the professed beliefs of many in science and industry, good science is not a commodity that courts can conveniently shop for in some extrasocietal marketplace of pure knowledge. There is no way for the law to access a domain of facts untouched by values or social interests. Scientific claims that are imported into the legal process are colored not only by the interests of the offering parties but also by the social, cultural, and political commitments of other actors in society: for example, the reluctance of experts to breach disciplinary solidarity, the law's desire to cloak morally difficult judgments with the "objective" authority of experts and instruments, and the public's demand for decisions that seem both open and rational. Historically, sociologically, and politically, the proposal that courts should increase their reliance on a value-neutral mainstream science is therefore extremely problematic.

Scientific closure and legal controversy do not, to begin with, stand in a predictably linear chronological relationship. Disagreement is endemic in science, and knowledge claims as often as not remain open-ended within the scientific community at times when they must be subjected to further testing in court. Scientific research is undertaken in many cases only after litigation pinpoints a possible causal connection. Where research exists, data still need to be aggregated and reanalyzed, sometimes through controversial scientific methods such as meta-analysis, in order to address the issues relevant to litigation. Context-specific information must be compiled to fill evidentiary gaps in cases ranging from employment discrimination to industrial and environmental disasters to patent infringement. Given its often limited relevance to scientific discovery, such information may never be independently reviewed or published, and publication if it occurs at all may postdate the needs of the legal process. Accordingly, the rule that courts should simply adopt the prevailing scientific opinion is often unworkable in practice. Consensus positions may develop around recurrent or widely distributed problems, but only after years of litigation and many interim efforts to produce definitive knowledge. Thus, it is not unusual that reliable epidemiological studies of breast implants were not available when Dow Corning settled the suits against it, or that clinical ecologists were formally read out of the medical mainstream only after they had begun to win some lawsuits.

At any given moment, moreover, the law's view of what constitutes the "mainstream" position in science may be an artifact of the legal system's limited and highly contingent ability to interrogate the scientific community. The fallibility of DNA typing, for example, was publicly demonstrated only after the technique had been deferred to in nearly two hundred trials. Not until the unscripted expert hearing in *People v. Castro* were there clear grounds for judges to question the methods by which laboratory tests had been conducted and interpreted. Problems in the representation of allele frequencies were slower to emerge. It took considerable expenditures of time and effort by the scientific and law enforcement communities, including the National Research Council and the Federal Bureau of Investigation, to produce a working consensus on standardized testing practices; as late as 1994 the "ceiling principle" endorsed by the NRC was still opposed by segments of the population genetics community.[1] Scientific consensus-building continually fell afoul of legal controversy because the law in effect created and structured the "credibility market" for DNA typing, from its initial transfer out of the laboratory to its deconstruction and partial reconstruction as a reliable forensic technique.

The production of scientific testimony for the courtroom is bound up with cross-cutting institutional and political imperatives that complicate the notion of science as a free-standing culture, independent of the law. The adversary process, the pressure to reach definite conclusions, and the selective use of knowledge by interested parties are only the most obvious social influences on the conversion of claims and observations into scientific evidence suitable for use in court. Less well attested but no less significant is the reorientation of scientific practices to suit the real or imagined needs of the legal system, from the initiation of strategic research to the adoption of unconventional modes of peer review and publication. Examples as varied as the Asilomar conferences on recombinant DNA research, the National Research Council study of DNA typing, the California Medical Association's report on clinical ecology, and the publication in *Science* of proposed evidentiary standards for use in toxic tort actions all testify to the scientific community's deep interest in constructing "mainstream" positions with an eye to their eventual adoption by the legal system. Science done (or interpreted) under these circumstances can scarcely be counted upon to maintain a persuasive impartiality when openly tested in court.[2]

Further, textbook science—the body of knowledge that is already in the public domain, having passed through science's critical filters—is rarely enough to satisfy the law's need for contextualized knowledge. In toxic tort cases, for example, even a large (though often undigested) body of literature on general causation typically will not answer the question of specific or individual causation upon which plaintiffs rest their claims for damages.[3] Mass torts, such as environmental disasters, require the production of site-specific knowledge, creating potential conflicts between knowledge considered valid by professional scientists and the knowledge compiled by local victims' groups.[4] Different types of evidence routinely elicit different credibility judgments from fact-finders: thus, epidemiological data may be favored over animal studies, a treating physician's testimony over a toxicologist's, a doctor's assessment of reasonableness over a patient's perceptions, and an instrumental reading over a police officer's or a coworker's reporting of individual experience. As noted in Chapter 3, such credibility judgments incorporate the fact-finder's own tacit understandings of science and expertise, although these private judgments may be hidden from critical review by rhetorically effective boundary work.[5] In this way, the fact-finder unavoidably enters into the construction of plausible facts in the courtroom as de facto participant observer.

If legally relevant knowledge is always interest-laden, then the choice between alternative scientific accounts necessarily involves normative, even political, judgments. Willingness to accept a particular knowledge claim amounts to an expression of confidence in the institutions and practices that produced it.[6] The researcher from the elite university, the representative of the powerful professional society, or the expert from the state agency may be judged as more authoritative than the "mere technician," who possesses neither Ph.D. nor publication record. In other contexts, the consultant employed by a corporate defendant or the university scientist whose work was supported by grants from industry may be dismissed as less credible than a disinterested-seeming witness with lesser professional credentials. Credibility judgments sometimes look systemic. In right-to-die cases, for example, medical views with regard to the intrusiveness or probable effectiveness of treatment options frequently overruled the patient's. In the Bendectin trials, juries reached more pro-plaintiff verdicts than judges. It appears that fact-finders' sensibilities with respect to causation may incorporate institutionalized social and political judgments, especially in products liability and medical malpractice cases.[7]

In sum, courts, like regulatory agencies, conduct the bulk of their scientific inquiries "at the frontiers of scientific knowledge," where claims are uncertain, contested, and fluid, rather than against a backdrop of largely settled "mainstream" knowledge. Instructing courts to take their cues from idealized stories of heroes and villains, such as the "good" scientific methodology of Gregor Mendel as opposed to the "bad" scientific practices of T. D. Lysenko, can therefore be deeply misleading.[8] Other forms of guidance are needed, more realistically attuned to the indeterminacy of scientific knowledge in the actual contexts of litigation, and mindful of the institutional strengths and weaknesses of judicial dispute resolution.

The Record of Judicial Accomplishment

How have legal institutions performed in facilitating the wide-ranging social readjustments that accompany, and indeed help define, scientific and technological change? Our responses, as I suggested in Chapter 1, can usefully be grouped under the headings of *deconstruction, civic education,* and *effectiveness.* These interlinked analytic criteria not only allow us to escape the conceptual straitjackets of "mainstream science" and "junk science" but are more compatible with the interactive and mutually constitutive nature of the relationship between science and technology and the legal process.

Deconstructing Expert Authority

The phrase "science at the bar" conjures up the image of a blustering, overconfident, perhaps inebriated science that has been called to account at the bar of justice. Indeed, the deconstruction of expert testimony by cross-examining attorneys is perhaps the most widely discussed aspect of science's relations with the law. But what does cross-examination really achieve? The credibility of science could in principle be assailed at many levels: specific scientific claims, their individual proponents, the methods and assumptions underlying the claims, and the institutions that certified those methods and assumptions. In reality, however, courtroom challenges to expertise are rather more selectively targeted, with the personal credibility of witnesses bearing the brunt of attack. Cross-examination, as Brian Wynne acutely observes, assumes that there is a substrate of "truth" to be precipitated out from "covert, extraneous bias (including values or opinions) or incompetence" in the presentation of testimony.[9] The law's institutional commitment to preserving the fact/value distinction (even if its artificiality is recognized by sophisticated practitioners) drives cross-examination toward an almost obsessive concern with inconsistencies in the witness's testimony and with biases, such as ties to economic interests, that are considered important in commonsense tests of credibility. The technical practices of lawyering thus shore up a deeper commitment to the notion of science as a reservoir of determinable facts. Science as a whole does not lie; it is only the occasional dishonest scientist on the witness stand who is culpable.

Not surprisingly, then, we find that the legal process exposes the cognitive and social commitments of individual expert witnesses more predictably than it identifies structurally or institutionally conditioned contingencies in scientific knowledge. The systematic problems of laboratory proficiency and unvalidated protocols in DNA typing went essentially unnoticed until qualified expert witnesses almost fortuitously called attention to the technique's methodological weaknesses. The law's entrenched deference to medical expertise obscured possibilities for conceptual development in such arenas as the

right to die, wrongful birth, and court-ordered obstetrical interventions. Communal assertions of expert authority carried
great weight in *Sterling v. Velsicol Chemical Corp.*, where the
court willingly adopted the boundary drawn by organized medicine between "real science" and clinical ecology.

At the same time, legal principles and arguments structure
the basic interpretive framework within which courts decide
which forms of institutionalized expertise they will defer to and
which not. As guardians of legal interpretation, courts control
the right to reopen and criticize legal categories and to formulate new ones, much as scientists do for scientific claims. But
decisions that look purely legal on the surface may determine
how far the deconstruction of science will be carried in particular cases. Perhaps the clearest illustration of this phenomenon was the "general acceptance" test, the conceptual ante-
cedent of today's "mainstream science," established in *Frye v.
United States*. The "general acceptance" standard gave courts
leeway for seventy years to decide for themselves whether
specific scientific claims were still open to challenge, while all
the time ostensibly deferring to science's independent authority.

Perhaps more subtly, the extension of "informed consent" in
Moore v. Regents of the University of California (discussed in
Chapter 5) protected traditional notions of property rights
against new possibilities suggested by science. The California
court could have created new property rights in human tissues
and cells following the "deconstruction" of the human body
through molecular research, but it chose not to exercise this
option. In *Diamond v. Chakrabarty*, the Supreme Court similarly passed up the opportunity to deconstruct the meanings of
invention and ownership under the Patent Act when presented
with the novel productions of biotechnology. Judicial boundary maintenance also came into play in *Anna J. v. Mark C.*, the
California gestational surrogacy case (discussed in Chapter 8),
in which the court refused to accept the recommendation of
the American College of Obstetrics and Gynecology that gestation be taken as the basis for motherhood. By characterizing
ACOG's conclusion as "law," the court effectively undermined
the professional society's authority to speak on this issue,

while protecting its own construction of genetics as the *scientifically* more appropriate marker of motherhood.[10]

Legal treatments of peer review again illustrate the finely differentiated interpretive strategies through which the law upholds the generic authority of science while preserving its own institutional right to contest scientific authority. The integrity of peer review has remained largely undisturbed when the practice is closely tied to the funding of research. By contrast, peer review and publication offer increasingly little protection against discovery requests by litigants. Here, in a context that directly impinges on judicial power, courts are prepared to let the parties look behind peer-reviewed scientific findings and contest their validity. Yet this strategy implicitly acknowledges the interpretive flexibility of raw data and concedes, if only tacitly, that peer review and publication do not provide uniquely authoritative interpretations of scientific observations. *Daubert v. Merrell Dow Pharmaceuticals, Inc.*, opted for a middle course between deference and distrust. The Supreme Court recognized the legitimating role of peer review in general but left open the possibility of challenging its results case by case.

The selectivity of legal deconstruction is equally apparent in cases involving technology. Controversies over technological risk and new developments in biomedicine display the positive power of litigation to challenge existing structures of technical expertise, identify emerging social norms, give them names and meaning, and sort out competing rights and obligations. Judicial ingenuity, however, may be exercised to inhibit as well as promote debate around new technologies. With respect to genetic engineering and biotechnology, for example, courts arguably construed their critical mandate too narrowly, thereby impeding the wider ethical and social debate that took place in some European legislatures. Courts did better at disclosing the interpretive flexibility of new reproductive technologies (who is a "mother," what counts as "privacy," can embryos be "owned"), aided perhaps by the decentralized nature of decisions and the possibility of different interpretations among and even within jurisdictions.

These examples underscore the fact that the deconstructive force of the law operates not only at the micro level of individual claims and controversies, but also at the macro level, where the law, as an institutionalized embodiment of distrust, challenges science's equally institutionalized claims to superior authority. Scientists, we know, are vigorous critics of each other's work, but this criticism is effective only within a basic envelope of trust that cannot be challenged without threatening the integrity of the entire scientific pursuit.[11] Decades of research on scientific controversies have documented the limited nature of the "organized skepticism" that the sociologist of science Robert Merton regarded as one of the constitutive norms of science.[12] With different institutional commitments, and with an equally powerful skeptical rhetoric at its command, the law can render transparent domains of contingency and constructedness in science that science's culturally bounded querying procedures could not have brought to light. Through repeated and incremental, if conflictual, interactions with science and technology, the legal system plays a vital part in exposing the presumptions of experts and holding them accountable to changing public values and expectations.

But what of the law's reflexive ability to deconstruct its own conceptions of science and expertise, especially when these conceptions are encoded within legal categories such as property, gender, causation, rationality, and, indeed, facts? Just as scientific peer review is constrained by the norms of science, so legal self-criticism, which proceeds through techniques of reasoning from and distinguishing among precedents, is constrained by the institutional commitments of the law. There are tantalizing hints from a wide range of cases that legal criticism only inconsistently opens up judicial understandings of science and technology. For example, in decisions on the admissibility of evidence *(Daubert)*, intellectual property *(Chakrabarty, Moore)*, creationism *(Edwards)*, toxic torts *(Agent Orange, In re Paoli)*, electromagnetic fields *(Criscuola)*, and surrogacy *(Baby M, Anna J.)*, judicial views about science and technology tacitly or explicitly warranted legal doctrines that, in turn, black-boxed or concealed the court's underlying conception of science. Most of all, the legal system jealously guards its power to declare what counts as science for purposes of the law. The extraordinarily undifferentiated and uncritical accounts of scientific methods and cultures found in much of the legal academic literature provide important intellectual support for this strategy.

Civic Education

At their most effective, legal proceedings have the capacity not only to bring to light the divergent technical understandings of experts but also to disclose their underlying normative and social commitments in ways that permit intelligent evaluation by lay persons. Adversary procedures, however, can be indiscriminately deconstructive in their impact and can obfuscate as well as advance critical inquiry. To what extent has litigation served to improve the quality of public debate on scientific and technological issues, whether by increasing participation, providing appropriate discursive and conceptual resources, or otherwise fostering deliberation?

Controversies about risk are perhaps the domain in which courts have made the most impressive contribution to the civic culture of American science and technology. Review by generalist judges symbolizes this nation's continued adherence to the principle that all governmental actions, however arcane or esoteric, must be explained in terms that are comprehensible to nonexpert audiences. By insisting on their prerogatives in this regard, courts have repeatedly affirmed that the ultimate power to guide technology policy is vested not in experts but in the citizenry. The aggressive substitution of judicial for administrative judgment under the rubric of the hard-look doctrine demonstrated the potential for abuse inherent in such a stance. As we saw in Chapters 4 and 7, judicial review also helped promote a kind of hyperrationality in the discourse of government: an overdependence on technical rationalization that made policymaking look superficially open and participatory but buried values beneath the veneer of objectivity.[13] Nonetheless, the overall effect of maintaining a dialogue between experts and the people, mediated by the legal process, appears to have been salutary. American attitudes toward science and technology remain generally optimistic, and the pervasive alienation from and mistrust of technology discernible in some segments of Western European polities to date have found no strong echoes in the United States.

Although courts have succeeded, sometimes brilliantly, in bringing to the surface public fears, concerns, and demands relating to technology, their record in transmitting these messages to other deliberative arenas remains equivocal. Right-to-die cases, as the Massachusetts Supreme Judicial Court observed in *Saikewicz* (discussed in Chapter 9), engaged the capacity of courts to provide "detached but passionate investigation and decision" on a matter of individual liberty. Courts were relatively successful over time both at conceptual clarification and at transferring policy responsibility to less adversarial institutions, such as ethics committees, professional societies, and legislatures. Other democratizing trends in judicial decisionmaking, however, such as the moves toward risk-spreading in tort law during the earlier part of the century and toward recognizing the autonomy of patients in the 1960s and 1970s, arguably reached a point of diminishing returns. Complex litigation introduced enormously important procedural innovations, but as an essentially bureaucratic move it had little useful impact on political discourse around issues of victim compensation. Scientists still decry the power of mass tort actions to lower the plaintiff's burden of proof, whereas experienced legal analysts question whether all the ingenuity invested in procedural reforms adequately benefits the victims for whom they were designed.[14]

If protracted controversy were synonymous with civic education, then the case of reproductive technologies would have marked the pinnacle of judicial achievement. Reality, of course, demands a more shaded verdict. As one key link in a sweeping realignment of scientific knowledge, material inventions, and social values, *Roe v. Wade* proved to be a vulnerable but unexpectedly durable legal construct. The decision neither satisfied legal critics (unlike *Diamond v. Chakrabarty* on patenting organisms) nor succeeded in disciplining an unruly network of social institutions and actors (unlike earlier Supreme Court decisions on contraception). *Roe*'s trimester framework and its unresolved doctrine of privacy succumbed to massive assaults from inside and outside the legal community. Yet some of *Roe*'s conceptual vocabulary enabled courts to decide other issues of reproductive rights, thereby almost imperceptibly strengthening *Roe*'s nonabsolutist balancing approach. And in twenty years the concept of fetal viability allowed a markedly more conservative Supreme Court to cut through the confusions of the antiabortion debate and give clearer definition to the central principle of women's reproductive liberty.

Effectiveness

Apart from deconstructing expert authority and providing another language and forum for political discourse, how effectively have courts answered to public demands for equity, efficiency, and responsiveness in decisions involving science and technology? Let us set aside to begin with the general criticisms of the tort system that occupy much of the academic literature, since these go beyond this book's primary focus on technological controversies. There is in any case only limited consensual knowledge about the tort system's social impacts, apart from an agreement that litigation in the American style is in sheer monetary terms an expensive habit.

The much-maligned inefficiency of the adjudicatory system, and the frequent second-fiddling by legislatures, assume a relatively benign cast when looked at through the lenses of federalism and pluralism. Modern technology raises wrenching questions about life and death, human nature and social relationships, to which twentieth-century America seems singularly reluctant to provide collective answers until multiple possible responses have been articulated in many discrete controversies. Given the religious, cultural, and ethical diversity of U.S. society, addressing claims at smaller units of discourse—the individual, the family, or the state—holds undeniable advantages. There is much to be gained in a pluralistic society by addressing value-laden technological disputes away from the glare and publicity of national legislation. Courts, for all their weaknesses as policymakers, possess certain offsetting virtues as mediators of conflicting values. The relatively decentralized, small-scale, and ad hoc character of judicial decisionmaking permits a more leisurely consideration of moral and ethical questions than is generally possible in the legislative arena.

Responsiveness is another of the legal system's great virtues, even though an overall assessment of judicial performance with regard to science and technology might fault the courts for participating too uncritically in the public steering of science and technology. Judicial fears about toxic chemicals and increased risk, coupled with judicial optimism about nuclear power and biotechnology, display much the same unreflective ambivalence that is revealed in public opinion surveys. Nonetheless, a fair evaluation of the judicial record counters the charge that courts by and large have dampened the development of science and technology. In cases ranging from patenting new life forms to regulating surrogacy, courts demonstrated their institutional capacity for breaking up policy stalemates and forcing legislatures to confront problems of growing public anxiety. Even where legislation came first, as in areas of environmental law, it fell to courts to supply the detailed and nuanced interpretive principles that the relatively blunt instrument of legislation could not provide. Few, if any, judicial decisions proved as restrictive for industry as the Delaney clause, the congressionally mandated ban on carcinogenic food additives that continued to influence public perceptions of risk nearly forty years after its enactment.

Policy Reform: Criticism with Credibility

However positively one views the democratizing influence of courts, one cannot ignore the widespread perception at this century's end that something in the American legal system's handling of science and technology is broken badly enough to need fixing. Proliferating proposals for improving judicial decisionmaking reveal at least three strands of thought about the nature of the problem and its possible solutions: (1) that courts should defer more to external sources of scientific authority ("mainstream science"); (2) that the legal system's established mechanisms for dealing with technical questions should be strengthened; and (3) that more alternatives to litigation should be sought, including litigation over scientific and technological issues. Each proposition calls for evaluation in the light of our overall argument that courts are places where science and technology are, and must necessarily be, constructed to serve both immediate and long-term social needs.

Separatist Schemes

Although many observers of the legal process see the worlds— or cultures—of law and science as separate, at least at the "core,"[15] the notion of a "science court"[16] or specialized branch of the judiciary with expertise in science and technology commands little political appeal. The historical commitment to generalist courts authorized by Article III of the Constitution and the legal system's hostility to the idea of "scientific separatism"[17] create formidable barriers against institutional bifurcation. Even the Court of Appeals for the Federal Circuit, whose jurisdiction over patent appeals makes it the most technical and specialized of the federal appellate courts, remains firmly tied to a holistic conception of the law.

Separatism, however, finds other expressions as well. One form is the injunction that judges should learn to "think like scientists" and should use specified, uniformly applicable criteria to determine whether the evidence before them is truly scientific. Despite their conceptual problems, the *Daubert* criteria for admissibility are likely to encourage further moves in this direction, even though my analysis suggests that attempts to reduce all scientific practice to a unitary, reductionist model will only confuse legal thinking and produce more uncertainty.

A less intrusive form of separatism is the preparation of handbooks, manuals, or panel reports that seek to systematize bodies of knowledge for use in resolving common types of scientific disputes. Such works are able to elucidate, without significant mythmaking, both substantive problems (such as toxic torts and DNA fingerprinting) and the processes by which science produces facts (such as replication and peer review). Thus, a reference manual on scientific evidence published in 1994 by the Federal Judicial Center instructs judges in how to recognize frequently litigated issues and reach better-informed decisions about them.[18] A section of the manual provides reference guides on seven areas of expert testimony: epidemiology, toxicology, survey research, forensic analysis of DNA, statistical inference, multiple regression analysis, and estimation of economic loss.

Apart from obsolescence, the main risk of this approach is that judges will fail to question the origins and foundations of the consensus that a manual or report purports to represent. What work was done to create the document? Does it make sense to accept its findings without further challenge (for example, through judicial notice of its conclusions)? The history of the first National Research Council study of DNA typing illustrates the reasons for concern. A panelist who participated in the study later reported that "the committee members had agreed to *let the report speak for itself* to avoid the emergence of conflicting gospels according to different members" (emphasis added).[19] A court that failed to look behind such possibly self-serving vows of silence could hardly meet its normative obligations, especially where, as in the case of DNA typing, guilt, innocence, and human lives might hang in the balance.

Finally, some scholars would like to connect the allegedly separate domains of law and science through new institutions rather than keep them distinct on the model of the science court. Bridging institutions or individuals such as "science counselors"[20] would be entrusted to carry information and critical perceptions back and forth across the perceived cultural divide. I have suggested that such proposals would simply substitute one process of social construction for another. Courts, as we have seen, are themselves a quintessential form of "bridging institution": they are places where scientists, lawyers, and lay persons participate in the production of legally relevant knowledge under particular, ritualized conventions for establishing credibility and authority. Why should one substitute for these tried-and-true "multicultural" institutions other, possibly less tested models of constructing knowledge and legal order?

A partial answer is that under certain circumstances nonadversarial institutions may serve both science and democratic values better than courts do. The objectives of criticism and civic education, for instance, are not always best entrusted to litigation. Scientific advisory bodies, in particular, have demonstrated their capacity to synthesize a common knowledge that satisfies norms of scientific, legal, and political accountability. At their best, such bodies can provide a "hard look" at the available evidence without falling prey to endless technical deconstruction. The Environmental Protection Agency, for example, fared better with the relatively nonadversarial Science Advisory Board, which permits negotiation among competing technical and moral sensibilities, than with the more courtlike Scientific Advisory Panel for the pesticides program, which sought to provide "scientific" answers to complex technical questions.[21] The more general point is that all institutions for knowledge production, whether courtlike or not, should be evaluated on their merits and in relation to the particular ends for which they were intended.

Training Judges and Informing Juries

Many methods have been proposed to strengthen the capacity of courts to evaluate scientific and technological controversies. The focus usually is on judges, because they are best positioned to help juries to weigh and contextualize scientific evidence, although the idea of blue-ribbon juries also deserves attention. Generally, these approaches hold promise because they combine flexibility with the potential for enhancing the fact-finder's awareness of the intertwined normative and technical issues in litigation.

The most radical scheme would require expert witnesses to be appointed by the court and answerable to the judge alone, as in European civil law jurisdictions. This prospect, however, holds little more than theoretical interest in a common law culture wedded to the virtues of party autonomy. Cross-cultural borrowings that seek to graft isolated procedural devices from one legal system onto another offer in any case only the slimmest hope of success; law, like language, is a system in which the elements are mutually interdependent in ways that only a "native speaker" can fully appreciate.[22] Moreover, federal judges have the power to appoint impartial expert witnesses under Rule 706 of the Federal Rules of Evidence, and few in the legal community believe that this authority needs to be expanded. Judges may be reluctant to invoke Rule 706 because of entrenched habit, lack of knowledge, and fear of usurping the jury's fact-finding prerogative; policies aimed at addressing these concerns may deserve further scrutiny.[23] Pretrial hearings offer another popular procedural device for increasing the give-and-take among scientific experts and broadening the range of expertise beyond the polarized extremes usually sought by the parties. As in the context of administrative rulemaking, hearings may create a more informative record than the more formal rituals of trial-type examination and cross-examination.

Although solutions such as these offer only incremental relief for what some see as a massive structural problem, there are strong reasons to favor incrementalism in the interactions between law, science, and technology. First, one of the greatest strengths of legal proceedings is precisely the ability to produce localized, context-specific epistemological and normative understandings that are not subordinated to inappropriately universal claims and standards. Second, we have seen that the deconstruction of expertise often happens most effectively through repeated encounters between scientists and lawyers, with the facts, the participating experts, and the legal rules of the game all changing from one disputing context to another. Both considerations would favor reforms that respect the diversity of problem-solving approaches currently represented in the American legal system.

Alternatives to Litigation

Trajectories of conflict such as those over abortion, fetal research, or the right to die point to the need for policymaking bodies that are less unwieldy than legislatures, as sensitive to values as courts, and yet better able to forge political compromises than a nonelected judiciary. In the United States, as in Britain, the blue-ribbon commission has been a favorite device for filling the institutional gap between courts and legislatures, but the effectiveness of such panels depends critically on the nature of their linkage to the larger political process. For example, Congress in 1988 created a bioethics committee to study the legal and ethical implications of the human genome project and to offer guidance on sensitive policy issues arising at the frontiers of the biomedical sciences. But the committee's operations were stalled when Congress proved unable to fill a vacancy created by the departure of Senator Lowell Weicker, a liberal Connecticut Republican who was defeated in the 1988 national election.[24] By contrast, a less political bioethics task force appointed in 1985 by Governor Mario Cuomo of New York successfully mediated among religious and ethnic groups and played a constructive role in political consensus-building.[25] Hospital ethics committees and the ethics advisory panels of scientific and technical societies can also serve as valuable adjuncts to the legal system. Free from the immediate pressures of litigation, such bodies are well positioned to carry out the dispassionate inquiry into professional standards and practices that so often eludes the reach of chronically overburdened courts.[26]

More generally, currents in the law that have little to do with science or technology may lead to reforms that will alleviate or radically redefine many of the problems noted in this book. Concerns about the complexity, cost, and inefficiency of litigation have been growing steadily for more than two decades, bringing far-reaching changes in the way law is practiced in America. The increasing acceptance of no-fault insurance schemes, negotiated rulemaking, complex litigation, pretrial disclosure, and methods of alternative dispute resolution testifies to a gradual movement away from the strictly adversarial approach to dealing with society's formal grievances. At the same time, increased monitoring of judicial performance and an emphasis on out-of-court settlements are beginning to expedite cases that do make their way into the legal arena. In the end, these macropolitical trends may do more to alleviate some of the grinding of gears between law, science, and technology than utopian schemes for marrying perfect rationality with perfect justice.

Conflict and Consensus in a Litigious Society

The law's dominion rests ultimately on its power to rebuild order and stability from uncertainty and chaos, and construction rather than conflict is therefore the appropriate note on which to close. A constructivist approach to the study of law, science, and technology offers distinctive insights into the reciprocal relationship of natural knowledge and social justice in American society. I therefore conclude my analysis with three examples of ways in which this approach could usefully be applied to deepen the agenda of legal research at the intersections of law, science, and technology. These specific examples should be seen as a more general invitation to diversify the methods, goals, and subjects of normal legal inquiry into conflicts dealing with science and technology.

First, the continuing national debate about the "litigation explosion" has been conducted to date largely through statistical claims and counterclaims concerning the phenomenon's "reality." A critical social history of this debate, focusing partly on the construction of "evidence" in support of divergent positions, would shed light on the persistence and inconclusiveness of the arguments. Studies incorporating perspectives from the sociology of knowledge would interestingly illuminate how legal as well as nonlegal actors construct their beliefs about science, expertise, and justice. The methods of controversy studies from the sociology of knowledge could productively be brought to bear on episodes such as the standoff that occurred in 1986 between the General Accounting Office (GAO) and the Justice Department over the extent of the litigation explosion in federal courts. An interagency task force headed by Justice reported that from 1974 through 1985 there had been a 758 percent increase in the number of products liability cases filed in federal courts. Asked to review these statistics by then Representative James Florio, a New Jersey Democrat, GAO countered that a disproportionate number of these filings could be attributed to a small handful of products (asbestos, Dalkon Shield, Bendectin). As could have been predicted in a politically charged controversy about the interpretation of science (in this case social science), GAO and Justice took issue with each other's baseline assumptions and methods of sampling and data collection.[27]

Second, the "literary technology" of the law should be more critically investigated in order to explicate its role in constructing dominant understandings of science and technology. The apparatus of legal scholarship focuses too narrowly on the "holdings" and formal "reasoning" that are thought to do the functional work of the law. Like the "junk DNA" of unknown utility that sits between our active genes, nonbinding statements of judicial opinion concerning science, technology, and the human condition have held relatively little interest for academics or professionals. Yet what passes as relatively meaningless in legal writing may be highly significant for purposes of analyzing judicial ideology and the cultural presumptions of the law. Particularly interesting are the metaphors and other rhetorical strategies through which courts establish working boundaries between the domains of fact and value, expertise and experience, knowledge and the law.

Finally, the rationales for public preferences for specific types of legal procedures and, more generally, the relationship between legal process, scientific authority, and political culture deserve more systematic investigation. A few empirical studies carried out in the 1970s provided suggestive if inconclusive evidence that the level of public satisfaction with the legal process rises when the participants "control the process of evidence presentation themselves while a third party controls the result." Such elements as "the absence of party confrontation and cross-examination of witnesses," these investigations concluded, would drive the legal process in directions that might diminish the participants' satisfaction.[28] Constructivist analyses of science, technology, and the law could provide a framework for probing the cultural and institutional foundations for such preferences. Studies of risk and regulation in similar Western societies, for example, have helped identify very different conceptions of what constitutes rationality and adequate democratic participation in the resolution of technical controversies.[29] In a time of accelerating social and technological uncertainty, such comparative and cross-disciplinary analyses promise to expand the repertoire of credible

public responses to changing knowledge and to reengage science and the law in mutually beneficial projects of reflection
and self-criticism.

Notes

Index

Notes

1. The Intersections of Science and Law

1. For a masterly account of the myths and images that shaped the U.S. public's attitudes toward nuclear weapons and nuclear energy, see Spencer Weart, *Nuclear Fear* (Cambridge, Mass.: Harvard University Press, 1988).

2. Rachel Carson, *Silent Spring* (Boston: Houghton Mifflin, 1962).

3. Humphrey Taylor, "Scientists, Doctors, and Teachers the Most Prestigious Occupations—But Doctors' and Lawyers' Prestige Falls Steeply," Harris Poll, June 7, 1992, in *American Public Opinion Data* (Louisville, Ky.: Opinion Research Service 1992), microfiche HAR, 7 June.

4. National Science Board (NSB), "Science and Technology: Public Attitudes and Public Understanding," in NSB, ed., *Science and Engineering Indicators—1993* (Washington, D.C.: U.S. Government Printing Office, 1993), pp. 204, 483.

5. Richard Topf, "Science, Public Policy, and the Authoritativeness of the Governmental Process," in Anthony Barker and Guy Peters, eds., *Expert Advice* (Pittsburgh: University of Pittsburgh Press, 1993), pp. 105–109.

6. Gina Kolata, "Forget the Butler; the Medical Industry Did It," *New York Times,* October 17, 1993, p. E3.

7. For an overview of these complaints, together with a reasoned rebuttal to some of the charges, see Marc Galanter, "Predators and Parasites: Lawyer-Bashing and Civil Justice," *Georgia Law Review* 28 (1994), 633–681.

8. Robert Gilpin and Christopher Wright, eds., *Scientists and National Policy-Making* (New York: Columbia University Press, 1964), p. 76.

9. The term "junk science" was popularized by Peter Huber, a lawyer and engineer, whose vivid though unscholarly indictment of the courts focuses primarily on their inability to discriminate between

marginal and mainstream science. See Huber, *Galileo's Revenge: Junk Science in the Courtroom* (New York: Basic Books, 1991).

10. Daniel E. Koshland, "Scientific Evidence in Court," *Science* 266 (1994), 1787.

11. See, for example, Steven Goldberg, "The Reluctant Embrace: Law and Science in America," *Georgetown Law Journal* 75 (1987), 1345.

12. Francisco J. Ayala and Bert Black, "Science and the Courts," *American Scientist* 81 (1993), 230–239; Peter H. Schuck, "Multi-Culturalism Redux: Science, Law, and Politics," *Yale Law and Policy Review* 11 (1993), 14–21.

13. For a sampling of the literature on the "culture clash" of law and science, see Philip M. Boffey, "Scientists and Bureaucrats: A Clash of Cultures on FDA Advisory Panel," *Science* 199 (1976), 1244–46; Leslie Roberts, "Science in Court: A Culture Clash," *Science* 257 (1992), 732–736; Steven Goldberg, *Culture Clash* (New York: New York University Press, 1994); and Schuck, "Multi-Culturalism Redux."

14. Ayala and Black, "Science and the Courts," p. 239.

15. Schuck, "Multi-Culturalism Redux," pp. 43–44; Goldberg, *Culture Clash*, pp. 103–108.

16. The classic work on paradigm changes in science is Thomas S. Kuhn, *The Structure of Scientific Revolutions* (Chicago: Chicago University Press, 1962).

17. Michael Polanyi, *Science, Faith and Society* (Oxford: Oxford University Press, 1946), pp. 45–46. Polanyi refers to law and science as regimes of "General Authority" because, in contrast to a regime of "Specific Authority" such as the Catholic church, these two disciplines decentralize the power to interpret rules. There is no official center to which individual discretionary decisions must be subjugated.

18. John Ziman, *Public Knowledge: An Essay concerning the Social Dimension of Science* (Cambridge: Cambridge University Press, 1968), pp. 14–15.

19. R. Austin Freeman, *The Eye of Osiris* (1911; reprint, New York: Carroll and Graf, 1986), pp. 123–124.

20. It has been suggested that one important difference between evidence in science and in the legal system is that in the latter "the information itself does not bear obvious credentials of its reliability and relevance." Accordingly, the law has designed "a highly developed 'law of evidence' for the presentation and testing of information offered as evidence in court cases." Jerome R. Ravetz, *Scientific Knowledge and Its Social Problems* (Oxford: Oxford University Press, 1971), p. 121. Recent work in the sociology of scientific knowledge questions

whether claims based on experimental observations do in fact bear "obvious credentials" of reliability. See, for instance, H. M. Collins, *Changing Order* (London: Sage, 1985). For the purposes of this book, the important point is not whether legal evidence is less obviously reliable than most scientific evidence, but rather whether the "law of evidence" is consistent with scientific criteria for assessing reliability.

21. Berry v. Chaplin, 74 Cal.2d 652 (1946).

22. Ayala and Black, "Science and the Courts," p. 230.

23. Michael J. Saks, "Accuracy v. Advocacy: Expert Testimony before the Bench," *Technology Review*, 90 (1987), 48.

24. Philip L. Bereano, "Courts as Institutions for Assessing Technology," in William A. Thomas, ed., *Scientists in the Legal System: Tolerated Meddlers or Essential Contributors?* (Ann Arbor: Ann Arbor Science, 1974), p. 85.

25. Lawrence M. Friedman, *The Republic of Choice: Law, Authority, and Culture* (Cambridge, Mass.: Harvard University Press, 1990).

26. Peter Huber, "Safety and the Second Best: The Hazards of Public Risk Management in the Courts," *Columbia Law Review* 85 (1985), 277–337.

27. Peter Huber, "Exorcists vs. Gatekeepers in Risk Regulation," *Regulation*, November/December 1983, pp. 23–32.

28. For a sampling of this literature, see Baruch Fischhoff et al., *Acceptable Risk* (Cambridge: Cambridge University Press, 1981); Mary Douglas and Aaron Wildavsky, *Risk and Culture* (Berkeley: University of California Press, 1982); Branden B. Johnson and Vincent Covello, *The Social and Cultural Construction of Risk* (Dordrecht: Reidel, 1987); Deborah G. Mayo and Rachelle D. Hollander, eds., *Acceptable Evidence: Science and Values in Risk Management* (New York: Oxford University Press, 1991).

29. Edmund W. Kitch, "The Vaccine Dilemma," *Issues in Science and Technology* 2 (1986), 108–121.

30. Philip M. Boffey, "Drug Shipments to Resume to Treat Rare Disorder," *New York Times*, November 6, 1986, p. B20.

31. Barbara J. Culliton, "Omnibus Health Bill: Vaccines, Drug Exports, Physician Peer Review," *Science* 234 (1986), 1313.

32. Patricia B. Gray, "Endless Trial," *Wall Street Journal*, January 13, 1987, p. 1.

33. Seth Mydans, "For Jurors, Facts Could Not Be Sifted from Fantasies," *New York Times*, January 19, 1990, p. A18.

34. Matthew L. Wald, "Jury in Cancer Death Suit Says Factory Polluted Wells," *New York Times*, July 29, 1986, p. A8.

35. Peter H. Schuck, *Agent Orange on Trial* (Cambridge, Mass.: Harvard University Press, 1986), pp. 263–265; Glenn Collins, "A To-

bacco Case's Legal Buccaneers," *New York Times*, March 6, 1995, p. D1.

36. James S. Kakalik et al., *Costs of Asbestos Litigation* (Santa Monica, Calif.: Rand Corporation, 1983).

37. See, for example, Michael J. Saks, "Do We Really Know Anything about the Behavior of the Tort Litigation System—and Why Not?" *University of Pennsylvania Law Review* 140 (1992), 1147–1292; Marc Galanter, "The Transnational Traffic in Legal Remedies," in Sheila Jasanoff, ed., *Learning from Disaster: Risk Management after Bhopal* (Philadelphia: University of Pennsylvania Press, 1994), pp. 135–144.

38. Litigation habits in Japan and Western Europe offer noteworthy contrasts with those in the United States.

39. Galanter, "Transnational Traffic in Legal Remedies," pp. 135–144.

40. Criscuola v. Power Authority of the State of New York, 81 N.Y.2d 649, 652 (1993).

41. 61 U.S.L.W. 4805, 113 S.Ct. 2786 (1993).

42. People v. Ojeda, 225 Cal. App. 3d 404 (1990).

43. Eric S. Lander and Bruce Budowle, "DNA Fingerprinting Dispute Laid to Rest," *Nature* 371 (1994), 735–738.

44. Association of American Physicians and Surgeons v. Hillary Rodham Clinton, et al., 1993 U.S. Dist. LEXIS 2597 (D.C.D.C. 1993).

45. Richard Stone, "Court Test for Plagiarism Detector?" *Science* 254 (1991), 1448.

46. "Ruling Left Intact in Sperm Bequeast," *New York Times*, September 5, 1993, p. 36; "Newlywed Hopes to Use Sperm of Dead Spouse to Start a Family," *New York Times*, June 5, 1994, p. 34; David W. Dunlap, "Sperm Donor Is Awarded Standing as Girl's Father," *New York Times*, November 19, 1994, p. 27; Ellen Goodman, "The Law vs. New Fact of Life," *Boston Globe*, January 26, 1995, p. 13.

47. "AIDS Victim's Colleagues Walk Out," *New York Times*, October 23, 1986, p. A24.

48. Gerry Elman, "Pasteur Institute Sues U.S. over AIDS Test Royalties," *Genetic Engineering News*, January 1986, p. 6.

49. Philip J. Hilts, "Americans Block French Move on AIDS Test," *New York Times*, September 20, 1992, p. 35; Jon Cohen, "U.S.-French Patent Dispute Heads for a Showdown," *Science* 265 (1994), 23–25.

50. Richard L. Madden, "Comatose Woman's Fetus Is Focus of Dispute," *New York Times*, March 8, 1987, p. 39.

51. Marcia Chambers, "Dead Baby's Mother Faces Criminal

Charges on Acts in Pregnancy," *New York Times,* October 9, 1986, p. A22.

52. Linda Greenhouse, "Court Order to Treat Baby Prompts a Debate on Ethics," *New York Times,* February 20, 1994, p. 12.

53. The interpretive flexibility of technological systems has been pointed out by scholars interested in exploring how technology comes to be stabilized in some ways and not in others. See Wiebe E. Bijker, Thomas P. Hughes, and Trevor Pinch, *The Social Construction of Technological Systems* (Cambridge, Mass.: MIT Press, 1987).

2. Changing Knowledge, Changing Rules

1. William Aldred's Case, 77 Eng. Rep. 817 (1610).

2. Barbara Ward and René Dubos, *Only One Earth: The Care and Maintenance of a Small Planet* (New York: Norton, 1972), p. 11.

3. Lawrence M. Friedman, *The Republic of Choice: Law, Authority, and Culture* (Cambridge, Mass.: Harvard University Press, 1990). For a study of the transnational legal and political impacts of technological change, see Sheila Jasanoff, ed., *Learning from Disaster: Risk Management after Bhopal* (Philadelphia: University of Pennsylvania Press, 1994).

4. Advertisement by AIG (American International Group), *Newsweek,* February 26, 1990, pp. 32–33. AIG's message draws heavily upon Peter Huber's ideas about liability, described and cited in note 5 below.

5. For a sweeping, at times intemperate, indictment of modern products liability law, see Peter Huber, *Liability* (New York: Basic Books, 1988). Huber argues that the consensual law of contracts, based on agreements between private parties, has given place to a public "law of coercion" that allows judges and juries to decide the safety obligations of manufacturers. Other writers, such as George Priest of Yale Law School, have called attention to the possible inefficiency of using tort law as an insurance mechanism geared toward risk spreading. Priest, "The New Legal Structure of Risk Control," *Daedalus* 119 (1991), 207–227. The best-documented connections between liability law and product unavailability are in the area of medical drugs and devices. See, for example, Leslie Roberts, "U.S. Lags on Birth Control Development," *Science* 247 (1990), 909 (describing a National Academy of Sciences study that calls for changes in liability law to bring new contraceptives on the market).

6. William L. Prosser, *Law of Torts* (St. Paul: West Publishing, 1971), p. 641.

7. 217 N.Y. 382 (1916).

8. G. Edward White, *Tort Law in America: An Intellectual History* (New York: Oxford University Press, 1985), p. 148.

9. Leon Green, "Tort Law Public Law in Disguise," *Texas Law Review* 38 (1959), 257.

10. Escola v. Coca-Cola Bottling Co., 24 Cal.2d 453, 462 (1944) (concurring opinion).

11. Greenman v. Yuba Power Products, Inc., 59 Cal.2d 57 (1963).

12. For example, the Supreme Court of New Jersey ruled in 1960 that a car manufacturer's potential liability extended to the purchaser's family members and others who were likely to use the product. Henningsen v. Bloomfield Motors, Inc., 32 N.J. 358 (1960).

13. Summers v. Tice, 33 Cal.2d 80 (1948).

14. 26 Cal.3d 588 (1980).

15. DES cases in other jurisdictions held that the determinate defendant requirement could be set aside as long as there was a rational formula for allocating liability within the group of potentially responsible manufacturers. See, for example, Bichler v. Eli Lilly and Co., 55 N.Y.2d 571 (1982); McElhaney v. Eli Lilly and Co., 564 F. Supp. 265 (1983); Collins v. Eli Lilly and Co., 342 N.W.2d 37 (Wis. 1984).

16. Peter S. Barth and H. Allan Hunt, *Workers' Compensation and Work-Related Illnesses and Diseases* (Cambridge, Mass.: MIT Press, 1980), p. 4.

17. 493 F.2d 1076 (5th Cir. 1973).

18. Deborah R. Hensler, William L. F. Felstiner, Molly Selvin, and Patricia A. Ebener, *Asbestos in the Courts: The Challenge of Mass Toxic Torts* (Santa Monica, Calif.: Rand Corporation, 1985), p. 20.

19. "Doctors in Boycott of Lawyer's Baby," *The Guardian,* May 15, 1986, p. 14.

20. General Accounting Office (GAO), *Medical Malpractice* (Washington, D.C., 1986), pp. 12–13.

21. Deborah Jones Merritt, "The Constitutional Balance between Health and Liberty," *Hastings Center Report,* December 1986, p. 3.

22. Duffield v. Williamsport School District, 162 Pa. 476 (1894).

23. New York State Ass'n for Retarded Children v. Carey, 466 F. Supp. 487 (E.D.N.Y. 1978).

24. Brune v. Belinkoff, 354 Mass. 102 (1968); Naccarato v. Grob, 384 Mich. 248 (1970); Pederson v. Dumonchel, 72 Wash.2d 73 (1967). See also Notes, *Stanford Law Review* 14 (1962), 884; and Notes, *Vanderbilt Law Review* 23 (1970), 729.

25. GAO, *Medical Malpractice,* p. 80.

26. Jeffrey O'Connell, *The Lawsuit Lottery* (New York: Free Press, 1979), chap. 1.

27. As of July 1985, nineteen states had standard-of-care provisions in effect. GAO, *Medical Malpractice*, p. 80.

28. Canterbury v. Spence, 464 F.2d 772 (D.C. Cir. 1972).

29. Jethro K. Lieberman, *The Litigious Society* (New York: Basic Books, 1981), p. 82.

30. Ibid., p. 88.

31. Helling v. Carey, 83 Wash.2d 514 (1974).

32. Lieberman, *Litigious Society*, p. 78.

33. Gates v. Jensen, 595 P.2d 919 (Wash. 1979).

34. GAO, *Medical Malpractice*, p. 18.

35. Fletcher v. Bealey, 28 Ch.D. 688–700 (1885).

36. For a history of the judicial role in the early years of NEPA litigation, see Frederick R. Anderson, *NEPA in the Courts* (Baltimore: Johns Hopkins University Press, 1973).

37. Christopher Stone, *Should Trees Have Standing? Toward Legal Rights for Natural Objects* (Los Altos, Calif.: William Kaufmann, 1974).

38. Reserve Mining Co. v. EPA, 514 F.2d 492 (8th Cir. 1975).

39. Ethyl Corp. v. EPA, 541 F.2d 1 (D.C. Cir. 1976).

40. 541 F.2d at 13.

41. 514 F.2d at 520.

42. 541 F.2d at 18.

43. 621 N.E.2d 1195 (N.Y. 1993).

44. 621 N.E.2d at 1196.

45. 62 U.S.L.W. 4576 (1994).

46. Linda Greenhouse, "High Court Limits the Public Power on Private Land," *New York Times*, June 25, 1994, p. 1.

3. The Law's Construction of Expertise

1. Quoted in Hubert W. Smith, "Scientific Proof," *Southern California Law Review* 16 (1943), 148.

2. See, for example, Lloyd L. Rosenthal, "The Development of the Use of Expert Testimony," *Law and Contemporary Problems* 2 (1935), 406–409.

3. Ibid., p. 410.

4. Folkes v. Chadd, 3 Doug. 157 (1782) (as quoted by Lawton, L. J. in *R. v. Turner*).

5. Smith, "Scientific Proof," p. 122.

6. Martin Shapiro, *Courts: A Comparative and Political Analysis* (Chicago: University of Chicago Press, 1981), p. 11.

7. A transcript of the pretrial hearing in the first U.S. case to use DNA fingerprinting as evidence, *State of Florida v. Tommie Lee An-*

drews, shows how important it was for prosecution experts to emphasize that this identification technique was the best available. Questions about the *absolute* reliability of the technique never got raised, even in cross-examination. Trial Proceedings, Information No. CR87–1400, Orange County Courthouse, Orlando, Fla., October 20, 1987.

8. John A. Jenkins, "Experts' Day in Court," *New York Times Magazine,* December 11, 1983, p. 98.

9. "Expert Witnesses: Booming Business for the Specialists," *New York Times,* July 5, 1987, p. 1.

10. See, for example, Calvin M. Kunin, "The Expert Witness in Medical Malpractice Litigation," *Annals of Internal Medicine* 100 (1984), 14.

11. Jenkins, "Experts' Day in Court," p. 105.

12. The practice is widely followed despite its possible pitfalls—as dramatized in the popular film *The Verdict,* in which a professional plaintiff's expert is severely embarrassed under cross-examination.

13. Peter Schuck, *Agent Orange on Trial* (Cambridge, Mass.: Harvard University Press, 1986), p. 230; Peter Huber, *Galileo's Revenge: Junk Science in the Courtroom* (New York: Basic Books, 199), pp. 96–100.

14. Joseph Sanders, "From Science to Evidence: The Testimony on Causation in the Bendectin Cases," *Stanford Law Review* 46 (1993), 36–47.

15. Lee Loevinger, "Law and Science as Rival Systems," *Jurimetrics* 8 (1966), 66.

16. Marvin E. Frankel, "The Search for Truth: An Umpireal View," *University of Pennsylvania Law Review* 123 (1975), 1036.

17. Ibid., p. 1038.

18. Peter Brett, "The Implications of Science for the Law," *McGill Law Journal* 18 (1972), 187.

19. Alan Usher, "The Expert Witness," *Medical Science Law* 25 (1985), 114.

20. Paul Meier, "Damned Liars and Expert Witnesses," *Journal of the American Statistical Association* 81 (1986), 273.

21. James E. Hough, "The Engineer as Expert Witness," *Civil Engineering ASCE,* December 1981, p. 57. This passage has evidently struck sympathetic chords in other professions, for it is repeated almost verbatim in Sanford M. Brown, "The Environmental Health Professional as an Expert Witness," *Journal of Environmental Health* 46 (1983), 86.

22. Hough, "The Engineer as Expert Witness," p. 58. These injunctions strikingly parallel the remarks of a former president of the Colorado Trial Lawyers Association, who likened the cross-examination

of an expert to playing a trout. William A. Trine, "Cross-Examining the Expert Witness in the Products Case," *Trial*, November 1983, p. 87. Interestingly, Hough, like Trine, used a fishing metaphor ("rise to the bait"). This language suggests that some technical professionals are prepared to "play" the cross-examining attorney just as trained litigators are inclined to "play" an uncooperative witness.

23. Brett, "The Implications of Science," pp. 186–187.

24. McLean v. Arkansas, 529 F. Supp. 1255 (E.D. Ark. 1982).

25. Philip L. Quinn, "The Philosopher of Science as Expert Witness," in James T. Cushing et al., eds., *Science and Reality: Recent Work in the Philosophy of Science* (Notre Dame: University of Notre Dame Press, 1984), p. 51. Whether intentionally or not, Quinn here was advocating something very close to the *Frye* rule, discussed below at note 62. The noted British philosopher Anthony Kenny took Quinn's suggestion a step further by proposing that experts should be permitted to testify only if their science meets threshold tests of consistency, method, cumulative ability, and predictiveness. Kenny recommended that proponents of a "new science" should seek something like a royal charter for their discipline before being allowed to testify as experts. Unlike that of *Frye*, Kenny's approach would make it impossible to negotiate credibility locally, within the context of specific cases. Anthony Kenny, "The Psychiatric Expert in Court," *Psychological Medicine* 14 (1984), 291–302.

26. Michael Ruse, "Commentary: The Academic as Expert Witness," *Science, Technology, and Human Values* 11 (1986), 72.

27. For an overview of problems that litigation poses to psychology and psychiatry, see Kenny, "The Psychiatric Expert in Court"; Kenneth F. Englade, "When Psychiatrists Take the Stand, Science Itself Goes on Trial," *The Scientist*, December 12, 1988, p. 1; Daniel Goleman, "Psychologists' Expert Testimony Called Unscientific," *New York Times*, October 11, 1988, p. C6; David Faust and Jay Ziskin, "The Expert Witness in Psychology and Psychiatry," *Science* 241 (1988), 31–35; Daniel E. Koshland, Jr., "Scientific Evidence in Court," *Science* 266 (1994), 1787.

28. See, for example, Meier, "Damned Liars and Expert Witnesses."

29. Marcia Angell, "Do Breast Implants Cause Systemic Disease?" *New England Journal of Medicine* 330 (1994), 1748. See also Sherine E. Gabriel et al., "Risk of Connective-Tissue Disease and Other Disorders after Breast Implantation," *New England Journal of Medicine* 330 (1994), 1697–1702.

30. George J. Annas, "Setting Standards for the Use of DNA-Typing Results in the Courtroom—the State of the Art," *New England Journal of Medicine* 326 (1992), 1643.

31. Huber, *Galileo's Revenge*; see, in particular, his account of the rise and fall of trauma-related cancer claims, pp. 39–56.

32. Eric S. Lander and Bruce Budowle, "DNA Fingerprinting Dispute Laid to Rest," *Nature* 371 (1994), 735.

33. Dorothy Nelkin and Laurence Tancredi, *Dangerous Diagnostics* (New York: Basic Books, 1989); Arielle Emmett, "Simulations on Trial," *Technology Review* 97 (1994), 30–36.

34. Sheila Jasanoff, *The Fifth Branch: Science Advisers as Policymakers* (Cambridge, Mass.: Harvard University Press, 1990), p. 68.

35. See, for example, Gina Kolata, "Two Chief Rivals in the Battle over DNA Evidence," *New York Times*, October 27, 1994, p. B14.

36. Brett, "Implications of Science for Law," p. 186 (quoting remark attributed to the eminent British colonial administrator and historian Lord Thomas Babington Macaulay).

37. Richardson v. Perales, 402 U.S. 413, 414 (1971).

38. Major works in this area include Thomas S. Kuhn, *The Structure of Scientific Revolutions* (Chicago: University of Chicago Press, 1962); Karin D. Knorr-Cetina and Michael Mulkay, eds., *Science Observed* (London; Sage, 1983); Bruno Latour and Steve Woolgar, *Laboratory Life* (Princeton: Princeton University Press, 1986); H. M. Collins, *Changing Order* (London: Sage, 1985); Bruno Latour, *Science in Action* (Cambridge, Mass.: Harvard University Press, 1987). For an application of these ideas to the specific context of legal expertise, see Roger Smith and Brian Wynne, eds., *Expert Evidence: Interpreting Science in the Law* (London: Routledge, 1989).

39. For an illuminating discussion of instruments and their embedding of conventions, see Latour, *Science in Action*, pp. 67–70.

40. I have described this phenomenon in detail in earlier writing. See, for example, Sheila Jasanoff, *Risk Management and Political Culture* (New York: Russell Sage Foundation, 1986); "The Problem of Rationality in American Health and Safety Regulation," in Smith and Wynne, *Expert Evidence*, pp. 151–183; and *The Fifth Branch*.

41. 603 F.2d 263 (2d Cir. 1979).

42. "The Law Tries to Decide Whether Whooping Cough Vaccine Causes Brain Damage: Professor Gordon Stewart Testifies," *British Medical Journal* 292 (1986), 1264–66.

43. 615 F. Supp. 262 (D. Ga. 1985).

44. 615 F. Supp. at 273.

45. 615 F. Supp. at 286, 291.

46. Robert K. Merton, "The Normative Structure of Science," reprinted in *The Sociology of Science* (Chicago: University of Chicago Press, 1973), pp. 267–278. In their seminal work on the culture of laboratory science, however, Bruno Latour and Steve Woolgar point

out that professional scientists do in practice frequently personalize their disputes over knowledge. Laboratory conversations about the work of fellow researchers illustrate the "common conflation of colleague and his substance: the credibility of the proposal and of the proposer are identical." Latour and Woolgar, *Laboratory Life*, p. 202.

47. See, for example, Sanders, "From Science to Evidence," pp. 39–41.

48. U.S. Congress, Office of Technology Assessment, *Genetic Witness: Forensic Uses of DNA Tests* (Washington, D.C.: U.S. Government Printing Office, 1990), p. 14.

49. 545 N.Y.S.2d 985 (Sup. 1989). See also Peter Banks, "Bench Notes," *Journal of NIH Research* 2 (1990), 75–77.

50. Colin Norman, "Maine Case Deals Blow to DNA Fingerprinting," *Science* 246 (1989), 1556–58.

51. For a more detailed exposition of these arguments, see Sheila Jasanoff, "What Judges Should Know about the Sociology of Science," *Jurimetrics* 32 (1992), 345–359. An effective deconstruction of the DNA fingerprinting technique, showing that it arguably lodges in seven different fields, is presented in William C. Thompson and Simon Ford, "DNA Typing: Acceptance and Weight of the New Genetic Identification Tests," *Virginia Law Review* 75 (1989), 45–108.

52. The "ceiling principle" made the worst-case assumption that the allele frequency at any locus should be taken as the maximum for any ethnic subpopulation. Multiplying these frequencies would then yield an upper bound for the frequency of a particular genotype in the population at large and yet produce odds suitable for obtaining criminal convictions. National Research Council, *DNA Technology in Forensic Science* (Washington, D.C.: National Academy Press, 1992); see also Lander and Budowle, "DNA Fingerprinting Dispute."

53. Letters from R. C. Lewontin and Daniel L. Hartl, *Nature*, 372 (1994), 398–399.

54. Milton R. Wessel, "Scientific Truth and the Courts," *Scientist*, March 9, 1987, p. 12.

55. R. E. Gots, "Medical Causation and Expert Testimony," *Regulatory Toxicology and Pharmacology* 6 (1986), 96–97.

56. Thompson v. Southern Pacific Transportation Co., 809 F.2d 1167 (5th Cir. 1987).

57. For a fuller account of Weinstein's reasoning in the summary judgment proceeding, see Schuck, *Agent Orange on Trial*, pp. 226–242.

58. See *In re* Agent Orange Product Liability Litigation, 818 F.2d 145 (2d Cir. 1987). Some courts have held that it is a reversible error for trial judges not to consult experts in relevant fields when deter-

mining whether an expert witness has reasonably relied on particular types of technical evidence. See, for example, *In re* Japanese Electronics Products, 723 F.2d 238 (3d Cir. 1983).

59. The California Supreme Court, too, applied the technician-versus-scientist distinction, but California's solution was to return the issue to the trial court for a better scientific record, an approach more frequently encountered in administrative than in adjudicatory decisionmaking. For a fuller analysis of these cases, see Sheila Jasanoff, "Judicial Construction of New Scientific Evidence," in Paul T. Durbin, ed., *Critical Perspectives in Nonacademic Science and Engineering* (Bethlehem, Pa.: Lehigh University Press, 1991), pp. 225–228.

60. People v. Ojeda, 225 Cal. App. 3d 404, 408 (1990).

61. 3 Cal. App. 4th 1326, 1333–34 (1992).

62. Frye v. United States, 293 F. 1013, 1014 (D.C. Cir. 1923).

63. See, for example, Philip Hiall Dixon, "*Frye* Standard of 'General Acceptance' for Admissibility of Scientific Evidence Rejected in Favor of Balancing Test," *Cornell Law Review* 64 (1979), 875–885; "Expert Testimony Based on Novel Scientific Techniques: Admissibility under the Federal Rules of Evidence," *George Washington Law Review* 48 (1980), 774–790; Mary W. Costley, "Scientific Evidence—Fryed to a Crisp," *South Texas Law Journal* 21 (1980), 62–79.

64. People v. Barbara, 400 Mich. 352, 405 (1977).

65. John W. Behringer, "Introduction to Proposals for a Model Rule on the Admissibility of Scientific Evidence," *Jurimetrics* 26 (1986), 238.

66. See, for example, Kenny, "The Psychiatric Expert," pp. 294–295.

67. United States v. Williams, 583 F.2d 1194 (2d Cir. 1978).

68. 113 S.Ct. 2786 (1993).

69. For a critique of the Court's ambivalence on this point, see Margaret G. Farrell, "*Daubert v. Merrell Dow Pharmaceuticals, Inc.*: Epistemology and Legal Process," *Cardozo Law Review* 15 (1994), 2183–2217.

70. Oliver Wendell Holmes, *The Common Law* (reprint, Boston: Little, Brown, 1963), p. 5: "The life of the law has not been logic: it has been experience."

71. Marcia Barinager, "Bendectin Case Dismissed," *Science* 267 (1995), 167.

72. Ron Simon, "High Court Throws Out Rigid Rules Excluding Scientific Evidence, Says Focus Must Be on Methods, Principles," Bureau of National Affairs, *Toxics Law Reporter* 8 (Summer/Fall 1993), 10.

73. Bert Black, Francisco J. Ayala, and Carol Saffran-Brinks, "Science and the Law in the Wake of *Daubert:* A New Search for Scientific Knowledge," *Texas Law Review* 72 (1994), 753, n. 260.

74. The science court proposal generated a substantial critical literature. See, for example, Arthur Kantrowitz, "Proposal for an Institution for Scientific Judgment," *Science* 156 (1967), 763–764; idem, "Controlling Technology Democratically," *American Scientist* 63 (1975), 505–509; "The Science Court Experiment: Criticisms and Responses," *Bulletin of the Atomic Scientists* 33 (1977), 44–50; Barry M. Casper, "Technology Policy and Democracy: Is the Proposed Science Court What We Need?" *Science* 194 (1976), 29–35. Also see articles in *RISK—Issues in Health and Safety* 4 (1993), a special issue on the science court.

75. Task Force of the Presidential Advisory Group on Anticipated Advances in Science and Technology, "The Science Court Experiment: An Interim Report," *Science* 193 (1976), 653.

76. Sheila Jasanoff and Dorothy Nelkin, "Science, Technology, and the Limits of Judicial Competence," *Science* 214 (1981), 1211–15.

77. Personal communication from Joseph S. Cecil, Division of Research, Federal Judicial Center, Washington, D.C., September 14, 1990.

78. John H. Langbein, "Restricting Adversary Involvement in the Proof of Fact: Lessons from Continental Civil Procedure" (paper presented at Cornell Law School, January 1985), p. 2. See also idem, "The German Advantage in Civil Procedure," *University of Chicago Law Review* 52 (1985), 823.

79. Pamela L. Johnston, Comment, *High Technology Law Journal* 2 (1988), 249.

80. Brian Wynne, *Rationality and Ritual: The Windscale Inquiry and Nuclear Decisions in Britain* (Chalfont St. Giles: British Society for the History of Science, 1982).

81. United Shoe Machinery Corp. v. United States, 110 F. Supp. 295 (D.Mass. 1953).

82. Harold Leventhal, "Environmental Decisionmaking and the Role of the Courts," *University of Pennsylvania Law Review* 122 (1974), 553.

83. Federal Judicial Center, *Reference Manual on Scientific Evidence* (Washington, D.C.: U.S. Government Printing Office, 1994).

4. The Technical Discourse of Government

1. Association of American Physicians and Surgeons v. Hillary Rodham Clinton, et al., 1993 U.S. Dist. LEXIS 2597 (D.D.C. 1993).

2. Comparative studies of regulatory policy have called attention to the special vulnerability of U.S. agencies to legal questioning. See, for example, Joseph L. Badaracco, Jr., *Loading the Dice* (Boston: Harvard Business School Press, 1985); Ronald Brickman, Sheila Jasanoff, and Thomas Ilgen, *Controlling Chemicals: The Politics of Regulation in Europe and the United States* (Ithaca: Cornell University Press, 1985); David Vogel, *National Styles of Regulation* (Ithaca: Cornell University Press, 1986).

3. Sheila Jasanoff, *Risk Management and Political Culture* (New York: Russell Sage Foundation, 1986).

4. Ted Greenwood, *Knowledge and Discretion in Government Regulation* (New York: Praeger, 1984).

5. For a comprehensive account of these problems, see Mark E. Rushefsky, *Making Cancer Policy* (Albany: SUNY Press, 1986).

6. Ethyl Corp. v. EPA, 541 F.2d 1, 67 (D.C. Cir. 1976).

7. The term "social regulation" was used by William Lilley III and James C. Miller III to describe the functions of many new regulatory agencies created in the 1970s. In contrast with their predecessors, which were concerned primarily with economic objectives, the new agencies were responsible for protecting public health, safety, and the environment. Their powers extended over a much greater diversity of businesses and industries, and they were authorized to enact significantly more economically and technologically burdensome regulations. See William Lilley III and James C. Miller III, "The New 'Social Regulation,'" *Public Interest*, no. 47 (Spring 1977), 49–61.

8. William H. Rodgers, Jr., "A Hard Look at Vermont Yankee: Environmental Law under Close Scrutiny," *Georgetown Law Journal* 67 (1979), 706.

9. 541 F.2d at 68.

10. 435 U.S. 519, 547 (1978).

11. For an argument that these developments were generally beneficial for EPA, see William Pedersen, "Formal Records and Informal Rulemaking," *Yale Law Journal* 85 (1975), 38.

12. Industrial Union Department, AFL-CIO v. Hodgson, 499 F.2d 467, 474 (D.C. Cir. 1974).

13. I have discussed the origins of the science policy paradigm at greater length in Sheila Jasanoff, *The Fifth Branch: Science Advisers as Policymakers* (Cambridge, Mass.: Harvard University Press, 1990), pp. 50–53.

14. 541 F.2d at 28.

15. Certified Color Manufacturers Association v. Mathews, 543 F.2d 284 (D.C. Cir. 1976); EDF v. EPA, 598 F.2d 62 (D.C. Cir. 1978);

Society of Plastics Industry, Inc. v. OSHA, 509 F.2d 1301 (2d Cir. 1975).

16. EDF v. EPA, 465 F.2d 528, 538 (D.C. Cir. 1972).

17. 543 F.2d at 297–298.

18. Hercules, Inc. v. EPA, 598 F.2d 91, 126 (D.C. Cir. 1978).

19. Lead Industries Association v. EPA, 647 F.2d 1130 (D.C. Cir. 1980).

20. R. Shep Melnick, *Regulation and the Courts* (Washington, D.C.: Brookings Institution, 1983), p. 356.

21. Specifically, the court suggested that the level of congressional concern conveyed by the Delaney clause justified special treatment of carcinogenic pesticides under the Federal Insecticide, Fungicide, and Rodenticide Act. EDF v. Ruckelshaus, 439 F.2d 584, 596, n. 41 (D.C. Cir. 1971).

22. Brickman, Jasanoff, and Ilgen, *Controlling Chemicals*, pp. 120–122.

23. In countries in which agencies were not subject to active judicial supervision there was no shift to a proactive, risk-based regulatory approach. See, for example, Brickman, Jasanoff, and Ilgen, *Controlling Chemicals;* Brendan Gillespie, Dave Eva, and Ron Johnston, "Carcinogenic Risk Assessment in the United States and Great Britain: The Case of Aldrin/Dieldrin," *Social Studies of Science* 9 (1979), 265–301. The failure to develop a risk-based approach is all the more remarkable because statutory safety standards governing agency action in some European countries seemed on their face more stringent than in the United States. Ronald Brickman and Sheila Jasanoff, "Concepts of Risk and Safety in Toxic Substances Regulation: A Comparison of France and the United States," *Policy Studies Journal* 9 (1980), 394–403.

24. Monsanto v. Kennedy, 613 F.2d 947 (D.C. Cir. 1979).

25. 613 F.2d at 955.

26. Melnick, *Regulation and the Courts*, p. 356.

27. American Petroleum Institute v. OSHA, 448 U.S. 607 (1980).

28. Section 6(b)(5) of the Occupational Safety and Health Act of 1970 provides that standards for toxic substances in the workplace should be set at the level that assures "to the extent feasible" that "no employee will suffer material impairment of health or functional capacity."

29. 448 U.S. at 653.

30. Kenneth S. Abraham and Richard A. Merrill, "Scientific Uncertainty in the Courts," *Issues in Science and Technology* 2 (1986), 98.

31. Devra Lee Davis, "The 'Shotgun Wedding' of Science and Law:

Risk Assessment and Judicial Review," *Columbia Journal of Environmental Law* 10 (1985), 81.

32. Ibid.

33. Thomas O. McGarity, "Beyond the Hard Look: A New Standard for Judicial review," *Natural Resources and Environment* 2 (1986), 66.

34. 62 U.S.L.W. 4576 (1994).

35. 701 F.2d 1137 (5th Cir. 1983).

36. See Davis, "Shotgun Wedding," p. 85; Richard Merrill, "The Legal System's Response to Scientific Uncertainty: The Role of Judicial Review," *Fundamental and Applied Toxicology* 4 (1984), S418–S425; Thomas O. McGarity, "Judicial Review of Scientific Rulemaking," *Science, Technology, and Human Values* 9 (1984), 97–106; Carl F. Cranor, *Regulating Toxic Substances: A Philosophy of Science and the Law* (New York: Oxford University Press, 1993), p. 122.

37. Abraham and Merrill, "Scientific Uncertainty in the Courts," p. 98.

38. Davis, "Shotgun Wedding," p. 85.

39. McGarity, "Beyond the Hard Look." McGarity suggested that courts might think of themselves as a "pass-fail professor" who fails the student "only when there is an inexcusable gap in the analysis, an obvious and significant misquote, or evidence of intellectual dishonesty." Ibid., p. 68.

40. 462 U.S. 87 (1983).

41. Vermont Yankee, 435 U.S. 519 (1978); Duke Power Co. v. Carolina Environmental Study Group, 438 U.S. 59 (1978).

42. 462 U.S. at 103.

43. 462 U.S. at 99.

44. 467 U.S. 837 (1984).

45. Peter H. Schuck and E. Donald Elliott, "To the *Chevron* Station: An Empirical Study of Federal Administrative Law," *Duke Law Journal*, 1990, pp. 984–1077.

46. See particularly Marc Roberts, Stephen Thomas, and Marc Landy, *The Environmental Protection Agency: Asking the Wrong Questions* (New York: Oxford University Press, 1990); and Sheila Jasanoff, "Science, Politics, and the Renegotiation of Expertise at EPA," *Osiris* 7 (1991), 195–217.

47. 824 F.2d 1146 (1987).

48. 831 F.2d 1108 (D.C. Cir. 1987).

49. Stephen Breyer, *Breaking the Vicious Circle: Toward Effective Risk Regulation* (Cambridge, Mass.: Harvard University Press, 1993).

50. For a more detailed discussion of this development, see Sheila Jasanoff, "Contested Boundaries in Policy-Relevant Science," *Social Studies of Science* 17 (1987), 217–218.

51. Jasanoff, "Science, Politics, and Renegotiation"; on congressional responses, see also Jasanoff, *The Fifth Branch*, pp. 89–90.

52. For more on this point, see Jasanoff, *The Fifth Branch,* pp. 240–241.

53. Asbestos Information Association v. OSHA, 727 F.2d 415 (5th Cir. 1984).

54. National Research Council, *Improving Risk Communication* (Washington, D.C.: National Academy Press, 1989), p. 21.

55. Comments of Patricia M. Wald in "The Contribution of the D.C. Circuit to Administrative Law" (transcript of a program presented at the ABA Section of Administrative Law), *Administrative Law Review* 40 (1988), 528.

5. Law in the Republic of Science

1. The different grounds on which science claims special status are well described in Bruce Bimber and David H. Guston, "Politics by the Same Means: Government and Science in the United States," in Sheila Jasanoff, Gerald E. Markle, James C. Petersen, and Trevor Pinch, eds., *Handbook of Science and Technology Studies* (Thousand Oaks, Calif.: Sage Publications, 1995), pp. 554–571.

2. For an overview of boundary-creating practices by scientists, see Thomas F. Gieryn, "Boundaries of Science," in Jasanoff, Markle, Petersen, and Pinch, *Handbook of Science and Technology Studies,* pp. 393–456; see also Gieryn, "Boundary-Work and the Demarcation of Science from Non-Science," *American Sociological Review* 48 (1983), 781–795.

3. Sociologists of science generally take a skeptical view of these claims. See, for example, Daryl E. Chubin and Edward J. Hackett, *Peerless Science: Peer Review and U.S. Science Policy* (Albany: SUNY Press, 1990).

4. "Duesberg Gets His Day in Court," *Science* 240 (1988), 279.

5. John Maddox, James Randi, and Walter W. Stewart, "'High-Dilution' Experiments a Delusion," *Nature* 334 (1988), 287–290.

6. Steven Goldberg, *Culture Clash: Law and Science in America* (New York: New York University Press, 1994), p. 61.

7. 411 F.2d 436 (2d Cir. 1969).

8. 411 F.2d at 443.

9. Apter v. Richardson, 510 F.2d 351 (7th Cir. 1975).

10. Steven Goldberg, "The Reluctant Embrace: Law and Science in America," *Georgetown Law Journal* 75 (1987), 1357.

11. 456 F. Supp. 1120 (S.D.N.Y. 1978).

12. Ujvarosy v. Sullivan, 1993 U.S. Dist. LEXIS 6330 (N.D. Cal. May 5, 1993).

13. 439 F.2d 584 (D.C. Cir. 1971).

14. 439 F.2d at 596, n. 41.

15. Lombardo v. Handler, 397 F. Supp. 792 (D.D.C. 1975), *aff'd*, 546 F.2d 1043 (D.C. Cir. 1976), *cert. denied.*, 431 U.S. 932 (1977).

16. The case related to the National Research Council's controversial report *Diet, Nutrition, and Cancer* (Washington, D.C.: National Academy Press, 1982). Thomas F. Howder, the Administrative Law Judge, accepted *in toto* the Academy's argument that "its 'confidential' policy regarding such materials is required for the performance of its essential functions. Were it not for the shielding of its deliberative and review process, participants in the Academy's studies would be inhibited in the candid exchange of views concerning often controversial scientific subjects. Such disclosures would have a chilling effect upon the conduct of vigorous internal debate and seriously impair the Academy's ability to produce reports of the best possible quality for the Government." In short, Howder agreed with the Academy that quality control in science is a matter best left to scientists and to the established processes of peer review. See Order Granting Motion to Limit Discovery Subpoena, In the Matter of General Nutrition, Inc., Federal Trade Commission Docket No. 9175, March 19, 1985.

17. For example, in *Wu v. National Endowment for the Humanities*, the Court of Appeals for the Fifth Circuit held that the Freedom of Information Act did not require disclosure of referee reports advising the Endowment about the desirability of funding a comprehensive history of China. The court accepted the argument that outside reviewers would be less candid if they knew their evaluations would be made public and concluded that the Endowment's interest in retaining the services of such experts outweighed the public's interest in disclosure. 460 F.2d 1030 (5th Cir. 1973), *cert. denied*, 410 U.S. 936 (1973).

18. University of Pennsylvania v. Equal Employment Opportunity Commission, 493 U.S. 182 (1989).

19. 113 S.Ct. 2786, 2797 (1993). The court cited work in the social studies of science (specifically, my own book on advisory committees) in support of this declaration. To be sure, trial courts looking for neat criteria of admissibility may still apply peer review as a rigid test. In principle, however, the court was unprepared to grant this degree of autonomy to scientists themselves.

20. In an unrelated but perhaps symptomatic event, the first direct challenge to the confidentiality of the National Science Foundation's

peer review system was filed under the Privacy Act in February 1994. See Eliot Marshall, "Researchers Sue to Get Reviewer Names," *Science* 263 (1994), 747.

21. Dow Chemical Company v. Allen, 672 F.2d 1262 (7th Cir. 1982) (preventing prepublication disclosure of herbicide toxicity study).

22. Richards of Rockford, Inc. v. Pacific Gas & Electric Co., 71 F.R.D. 388 (N.D.Cal. 1976) (names of company employees could be protected in study of defective cooling equipment at California power plant); Farnsworth v. Proctor and Gamble Co., 758 F.2d 1545 (11th Cir. 1985) (confidentiality of sources can be maintained in litigation involving toxic shock syndrome).

23. Eliot Marshall, "Court Orders 'Sharing' of Data," *Science* 261 (1993), 284–286. See also *In re* American Tobacco Company, 880 F.2d 1520 (2d Cir. 1989), and Wright v. Jeep Corp., 547 F. Supp. 871 (E.D. Mich. 1982).

24. See William Broad and Nicholas Wade, *Betrayers of the Truth* (Oxford: Oxford University Press, 1982); Alexander Kohn, *False Prophets: Fraud and Error in Science and Medicine* (Oxford: Basil Blackwell, 1986). For a partial listing of the innumerable newspaper and journal reports, editorials, and letters written about scientific misconduct in the 1980s, see Marcel C. LaFollette, "Ethical Misconduct in Research Communication: An Annotated Bibliography" (photocopy, Massachusetts Institute of Technology, August 1988). Further, see Walter W. Stewart and Ned Feder, "The Integrity of the Scientific Literature," *Nature* 325 (1987), 207–214. For an extended narrative of the so-called Baltimore affair, see Chubin and Hackett, *Peerless Science,* pp. 138–153.

25. Robert M. Andersen, "The Federal Government's Role in Regulating Misconduct in Scientific and Technological Research," *Journal of Law and Technology* 3 (1988), 121–148. The Public Health Service's proposed rules were published at *Federal Register* 53, September 19, 1988, pp. 36345–50.

26. 756 F. Supp. 1172 (W.D. Wis. 1990)

27. Christopher Anderson, "Popovic Is Cleared on All Charges; Gallo Case in Doubt," *Science* 262 (1993), 981–983; "ORI Drops Gallo Case in Legal Dispute," *Science* 262 (1993), 1202–03.

28. 648 F. Supp. 1248 (S.D.N.Y. 1988); 868 F.2d 1313 (2d Cir. 1989), *cert. denied,* 110 S.Ct. 219 (1989).

29. In a passage remarkable for its lack of independent critical acumen, the district court stated: "it was the defendant who opened the doors for Dr. Weissman, making all of her research and writing possible and professionally recognized. The defendant was the one who had acquired the radiopharmaceutical from the manufacturer as a

result of his renown in the medical community . . . Even when the plaintiff was able to obtain a radiopharmaceutical in her own name, the defendant was ultimately responsible for the use of the drugs, as the person with whom 'the buck stops.'" Weissman v. Freeman, 648 F. Supp. 1248 at 1259.

30. Alexander M. Capron, "Human Experimentation," *Biolaw* (1986), 227–228. See also Henry K. Beecher, "Ethics and Clinical Research," *New England Journal of Medicine* 274 (1966), 1354–60. As recently as 1993, the U.S. Department of Energy disclosed records of radiation exposure experiments conducted on civilians without their consent. Particularly disturbing was a study done at a Massachusetts school for the retarded in which children were fed radiation-contaminated food as part of their prescribed diet.

31. Arguably, the first appellate court decision involving human experimentation was handed down in a 1776 English case in which a surgeon was held liable for using a novel technique to set a broken leg. The court said, *inter alia,* that the surgeon should have informed the patient in advance about what he planned to do. See George Annas, Leonard Glantz, and Barbara Katz, *Informed Consent to Human Experimentation: The Subject's Dilemma* (Cambridge, Mass.: Ballinger, 1977), pp. 2–3.

32. An early American case that distinguished between quackery and legitimate experimentation was Baldor v. Rogers, 81 So.2d 658 (Fla. 1955). The court concluded that medical experimentation was not automatically to be regarded as malpractice and should even be encouraged when no effective treatment was otherwise available.

33. Maria Woltjen, "Regulation of Informed Consent to Human Experimentation," *Loyola University Law Journal* 17 (1986), 510.

34. Regulations issued pursuant to the National Research Act of 1974 set forth consent requirements that must be satisfied in all research funded by the extramural programs of the Department of Health and Human Services (DHHS; formerly Health, Education and Welfare). FDA adopted similar regulations for clinical studies carried out to obtain approvals for drugs and medical devices. A few states, notably New York and California, enacted their own informed-consent statutes for biomedical research. Such laws aimed to fill gaps in federal regulation and were generally consistent with the approach taken by DHHS. Woltjen, "Regulation of Informed Consent," pp. 518–523.

35. U.S. Congress, Office of Technology Assessment, *Human Gene Therapy* (Washington, D.C., 1984), p. 45.

36. Peter H. Schuck, "Rethinking Informed Consent," *Yale Law Journal* 103 (1994), 899–959.

37. See, for example, Karp v. Cooley, 493 F.2d 408 (5th Cir. 1974) at 419, n. 11.

38. Canterbury v. Spence, 464 F.2d 772 (D.C. Cir. 1972), *cert. denied*, 409 U.S. 1064 (1972).

39. 23 Cal. Rptr. 2d 131 (1993).

40. Annas, Glantz, and Katz, *Informed Consent,* pp. 32–33.

41. Halushka v. University of Saskatchewan, 53 D.L.R.2d 436 (Sask. 1965). The subject in this case underwent a procedure that resulted in a temporary cessation of his heartbeat, serious emergency surgery, and fourteen days of hospitalization. The court held that he was inadequately informed of the procedure's risks and that, under the circumstances, a signed consent form did not absolve the researchers of their liability.

42. Annas, Glantz, and Katz, *Informed Consent,* p. 12.

43. 493 F.2d 408 (5th Cir. 1974).

44. The *Karp* court did not consider the adequacy of Cooley's protocol for using the artificial heart. It seems clear that the procedure would not have satisfied the criteria issued by the National Heart and Lung Institute in 1974 for human testing of therapeutic devices. For example, Cooley's surgical procedure was neither adequately pretested on animals nor formally peer reviewed. Annas, Glantz, and Katz, *Informed Consent,* p. 13. Given the limited nature of the evidence before it, however, it is not surprising that the *Karp* court failed to address these aspects of the operation. The most one can deduce from the case is that courts are unlikely to analyze the validity of a scientific procedure with any sophistication unless they are alerted to this issue by the litigants and given explicit guidance by the scientific community.

45. 493 F.2d at 423.

46. George J. Annas, "Changing the Consent Rules for Desert Storm," *New England Journal of Medicine* 326 (1992), 770–773.

47. 452 N.Y.S.2d 875 (App. Div. 1982).

48. 452 N.Y.S.2d at 879.

49. Mink v. University of Chicago, 460 F. Supp. 713 (1978).

50. Mark Crawford, "Court Rules Cells Are the Patient's Property," *Science* 241 (1988), 653.

51. John Moore v. Regents of the University of California et al., 241 Cal. Rptr. 147 (1990) at 149.

52. James Boyle, "A Theory of Law and Information: Copyright, Spleens, Blackmail, and Insider Trading," *California Law Review* 80 (1992), 1413–1540; see especially pp. 1429–32.

53. Keith Schneider, "Theft of Infected Cats from U.S. Lab Spurs Alert," *New York Times*, August 25, 1987, p. A14.

54. Constance Holden, "Centers Targeted by Activists," *Science* 232 (1986), 149.

55. Schneider, "Theft of Infected Cats."

56. Dianne Dumanoski, "The Animal-Rights Underground," *Boston Globe Magazine,* April 22, 1987, p. 17.

57. Violence by the antiabortion movement may have begun to harden judicial opposition by the early 1990s. See, for example, the 1994 decision in which the Supreme Court extended federal racketeering laws to cover the activities of antiabortion groups across state lines.

58. Animal Lovers Volunteer Association (ALVA) v. Weinberger, 765 F.2d 937 (9th Cir. 1985). See also Humane Society of the United States v. Block, Civil Action No. 81-2691 (D.D.C. 1982); Fund for Animals v. Malone, Civil Action No. 81-2977 (D.D.C. 1982); International Primate Protection League v. Institute for Behavioral Research, 799 F.2d 934 (4th Cir. 1986).

59. Taub v. State of Maryland, 463 A.2d 819 (Md. 1983).

60. 799 F.2d at 935.

61. 799 F.2d at 939.

62. Scopes v. State of Tennessee, 154 Tenn. 105 (1927).

63. Ronald L. Numbers, "Creationism in 20th-Century America," *Science* 218 (1982), 538–544. See also idem, *The Creationists* (New York: Knopf, 1992).

64. Epperson v. Arkansas, 393 U.S. 97 (1968).

65. 482 U.S. 578 (1987).

66. Brief for *Amicus curiae* the National Academy of Sciences Urging Affirmance in *Edwards v. Aguillard,* No. 85-1513, Supreme Court, October Term, 1985, pp. 6–16.

67. 403 U.S. 602 (1971).

68. 482 U.S. at 591.

69. Goldberg, *Culture Clash,* pp. 78–79.

70. After initially succeeding at the district court level, this argument, too, was rejected by a federal court of appeals. See Smith v. Board of School Commissioners of Mobile County, 635 F. Supp. 939 (S.D. Ala. 1987); *reversed,* Smith v. Board of School Commissioners of Mobile County, 827 F.2d 684 (11th Cir. 1987).

71. For some characteristic expressions of these views, see Gary Taubes, "Misconduct: Views from the Trenches," *Science* 261 (1993), 1108–11.

6. Toxic Torts and the Politics of Causation

1. For a sampling of such opinion, see Kenneth R. Foster, David E. Bernstein, and Peter W. Huber, "Science and the Toxic Tort," *Science* 261 (1993), 1509–10; Stephen D. Sugarman, "The Need to Reform

Personal Injury Law Leaving Scientific Disputes to Scientists," *Science* 248 (1990), 823–827; Peter W. Huber, *Galileo's Revenge: Junk Science in the Courtroom* (New York: Basic Books, 1991). See also the Republican Party's "common sense legal reforms" in the "Contract with America," Washington, D.C., September 1994. The technocratic bias of these proposals is also reflected in some contemporary writing on regulation; see Stephen Breyer, *Breaking the Vicious Circle: Toward Effective Risk Regulation* (Cambridge, Mass.: Harvard University Press, 1993).

2. See, for example, David Rosenberg, "The Causal Connection in Mass Exposure Cases: A 'Public Law' Vision of the Tort System," *Harvard Law Review* 97 (1984), 851–929. Briefly, Rosenberg urged that damages be awarded according to a rule of "proportional liability," under which defendants would be liable only for the percentage of the plaintiffs' injuries that could be attributed to the defendants' activities. The Rosenberg proposal also contemplated greater judicial use of public law mechanisms such as class actions, damage schedules, and insurance funds established by payments from defendants. For an indication of the proactive energy that lawyers bring to mass tort actions, see Glenn Collins, "A Tobacco Case's Legal Buccaneers," *New York Times,* March 6, 1995, p. D1.

3. Michael J. Saks, "Do We Really Know Anything about the Behavior of the Tort Litigation System—and Why Not?" *University of Pennsylvania Law Review* 140 (1992), 1147–1292.

4. Mary Douglas and Aaron Wildavsky, *Risk and Culture* (Berkeley: University of California Press, 1982), pp. 31–32.

5. Lawrence M. Friedman, *The Republic of Choice: Law, Authority, and Culture* (Cambridge, Mass.: Harvard University Press, 1990), p. 60.

6. For a more extensive description of these impacts, see *Man's Impact on the Global Environment,* Report of the Study of Critical Environmental Problems (SCEP) (Cambridge, Mass.: MIT Press, 1970), pp. 126–136.

7. The evidence purporting to show that DDT is carcinogenic in rats and mice has been attacked on a variety of methodological grounds. For one such critique, see Edith Efron, *The Apocalyptics* (New York: Simon and Schuster, 1984), pp. 267–270. As Efron indicates, the interpretation of the animal data on DDT involves numerous subjective judgments, and the question of the pesticide's carcinogenicity thus remains controversial to this day.

8. Association of Trial Lawyers of America, *Toxic Torts* (Washington, D.C., 1977).

9. William R. Ginsberg and Lois Weiss, "Common Law Liability for

Toxic Torts: A Phantom Remedy," *Hofstra Law Review* 9 (1980–81), 859–941. Pointing to the difficulty of proving causation, the authors recommended the creation of an administrative remedy to compensate victims of hazardous waste contamination incidents such as Love Canal.

10. Marcia Angell, "Do Breast Implants Cause Systemic Disease?" *New England Journal of Medicine* 330 (1994), 1748–49. See also Chapter 3.

11. *In re* Paoli Railroad Yard PCB Litigation, 916 F.2d 829 (3d Cir. 1990).

12. 939 F.2d 1106 (5th Cir. 1991).

13. There have been no systematic studies of the role played by treating physicians in the legal process. There are, however, some comparative data suggesting that treating physicians may enjoy a more privileged status in the United States than in other legal systems. Specifically, in U.S. asbestos lawsuits, the injured worker is generally represented by his own physician, whereas in British compensation cases the experts are generally drawn from a more select community of specialists in dust-related diseases. Personal communication from Robert Dingwall, Centre for Socio-Legal Studies, Oxford, England.

14. 552 F. Supp. 1293 (D.D.C. 1982).

15. 552 F. Supp. at 1300.

16. In some jurisdictions this favored treatment has been elevated to the status of a legal presumption. Menendez v. Continental Insurance Co., 515 So.2d 525 (La. App. 1st Cir. 1987).

17. Civil Action No. 84-3235 (D.N.J. April 16, 1986).

18. Ibid., p. 6.

19. 809 F.2d 1167 (5th Cir. 1987).

20. See, in particular, Weinstein's dismissal of testimony by Dr. Samuel Epstein: "An extensive deposition of Dr. Epstein . . . consists of a devastatingly successful showing of his lack of knowledge of the medical and other background of those on whose behalf he submitted affidavits." 611 F. Supp. at 1238.

21. 615 F. Supp. 262 (D.Ga. 1985).

22. Jethro K. Lieberman, *The Litigious Society* (New York: Basic Books, 1981), p. 82.

23. David H. Kaye, "On Standards and Sociology," *Jurimetrics* 32 (1992), 545.

24. 588 F. Supp. 247 (1984).

25. To satisfy the "substantial factor" test, the plaintiffs had to show at a minimum (1) probable exposure to radiation levels in excess of "background" rates, (2) injuries of a type consistent with expo-

sure to ionizing radiation, and (3) residency near test site during period of atmospheric testing. 588 F. Supp. at 428.

26. See Sheila L. Birnbaum, "Remarks on Expert Testimony, Rules of Evidence, and Judicial Interpretation," in Institute for Health Policy Analysis, *Causation and Financial Compensation,* Conference Proceedings (Washington, D.C., 1986).

27. Allen v. U.S., 816 F.2d 1417 (10th Cir. 1987).

28. *In re* Agent Orange Produce Liability Litigation, 611 F. Supp. 1223 (D.N.Y. 1985).

29. 611 F. Supp. at 1231.

30. Foster, Bernstein, and Huber, "Science and the Toxic Tort," p. 1509.

31. Sheila Jasanoff, *The Fifth Branch: Science Advisers as Policymakers* (Cambridge, Mass.: Harvard University Press, 1990), pp. 145–146.

32. Dorothy Nelkin and Laurence Tancredi, *Dangerous Diagnostics* (New York: Basic Books, 1989).

33. Clifford Zatz, unpublished remarks presented at Cornell University, Institute for Comparative and Environmental Toxicology, Symposium on "Immunotoxicology: From Lab to Law," October 1987.

34. 525 A.2d 287 (N.J. 1987).

35. 525 A.2d at 309.

36. 647 F. Supp. 303 (D.Tenn. 1986).

37. 855 F.2d 1188, 1205 (6th Cir. 1988).

38. *In re* Paoli Railroad Yard PCB Litigation, 916 F.2d at 852.

39. 621 N.E.2d 1195 (N.Y. 1993).

40. See Ulrich Beck, *The Risk Society: Towards a New Modernity* (London: Sage Publications, 1992), for one especially cogent exposition of these views; astonishingly but significantly, the book was a runaway bestseller in Germany, where over 250,000 copies of the German original were sold.

41. John W. Gulliver and Christine C. Vito, "EMF and Transmission Line Siting: The Emerging State Regulatory Framework and Implications for Utilities," *Natural Resources and Environment* 7 (Winter 1993), 12–15.

42. 460 U.S. 766 (1983).

43. See, for example, Foster, Bernstein, and Huber, "Science and the Toxic Tort"; and Huber, *Galileo's Revenge.* The term "mainstream science" entered the national political discourse through the efforts of Vice President Dan Quayle's Competitiveness Council, a conservative policy lobby based in the White House during the Bush years. See President's Council on Competitiveness, *Agenda for Civil Justice Reform* (Washington, D.C.: U.S. Government Printing Office, 1991).

Subsequently the issue was taken up by the Republican Party in its 1994 "Contract With America."

44. Robert Reinhold, "When Life Is Toxic," *New York Times Magazine,* September 16, 1990, p. 51.

45. 628 F. Supp. 1219 (D.Mass. 1986).

46. Laurie A. Rich, "'No Winners' in an $8–9 Million Settlement," *Chemical Week,* October 8, 1986, pp. 18–19.

47. 515 So.2d 525 (La. App. 1st Cir. 1987).

48. 515 So.2d at 527.

49. Civil Action No. 3-84-0219-H (N.D. Tex. 1985).

50. 474 So.2d 1320 (La. App. 3rd Cir. 1985). Cited in Peter N. Sheridan and Bradley S. Tupi, "Joint Arrangement Results in Victory for Chemical and Insurance Companies," 332 PLI/Lit 447 (1987).

51. 855 F.2d 1188 (6th Cir. 1988).

52. American Academy of Allergy and Immunology, "Position Statements: Clinical Ecology," *Journal of Allergy and Clinical Immunology* 78 (1986), 270.

53. California Medical Association Scientific Board Task Force on Clinical Ecology, "Clinical Ecology—A Critical Appraisal," *Western Journal of Medicine* 144 (1986), 240.

54. Ibid., p. 243.

55. 855 F.2d at 1208.

56. Saks, "Do We Really Know Anything?" p. 1287. See also James A. Henderson Jr. and Theodore Eisenberg, "The Quiet Revolution in Products Liability: An Empirical Study of Legal Change," *UCLA Law Review* 37 (1990), 479–553.

57. See Toxic Tort Act, H.R. 1049, 96th Cong., 1st sess. (1979); Toxic Waste and Tort Act of 1979, H.R. 3797, 96th Cong., 1st sess. (1979); Hazardous Waste Control and Toxic Tort Act of 1979, H.R. 5291, 96th Cong., 1st sess. (1979). For a contemporaneous analysis of the pros and cons of administrative compensation, see Stephen M. Soble, "A Proposal for the Administrative Compensation of Victims of Toxic Substance Pollution: A Model Act," *Harvard Journal on Legislation* 14 (1977), 683–824.

58. *Injuries and Damages from Hazardous Wastes: Analysis and Improvement of Legal Remedies, Report to Congress in Compliance with Section 301(e) of the Comprehensive Environmental Response, Compensation, and Liability Act of 1980 (P.L. 96-510) by the "Superfund Section 301(e) Study Group,"* 97th Cong., 2nd sess. (1982).

59. House Committee on Public Works and Transportation, Subcommittee on Investigations and Oversight, *Hazardous Waste Contamination of Water Resources (Compensation of Victims Exposed to Hazardous Wastes),* 98th Cong., 1st sess. (1983).

60. House Committee on Public Works and Transportation, Subcommittee on Investigations and Oversight, *Hazardous Waste Exposure Victims Compensation (Assessing Risks from Ambient, Nonworkplace Exposure, and the Need for Additional Remedies)*, 99th Cong., 2nd sess. (1986), p. 27.

61. Saks, "Do We Really Know Anything?" p. 1289.

62. Tom Durkin and William L. F. Felstiner, "Bad Arithmetic: Disaster Litigation as Less than the Sum of Its Parts," in Sheila Jasanoff, ed., *Learning from Disaster: Risk Management after Bhopal* (Philadelphia: University of Pennsylvania Press, 1994), pp. 158–179.

7. Legal Encounters with Genetic Engineering

1. Christopher Walters, *American Way*, July 1, 1991, p. 73. That Rifkin rated an interview in the American Airlines magazine was in itself a token of his unique political status.

2. "A Novel Strain of Recklessness," *New York Times*, April 6, 1986, sec. 4, p. 22.

3. "Rifkin against the World," *Los Angeles Times*, April 17, 1986, pt. II, p. 6.

4. "Ban Experiments in Genetic Engineering?" *U.S. News & World Report*, October 8, 1984, p. 44.

5. Daniel E. Koshland, "Judicial Impact Statements," *Science* 239 (1988), 1225.

6. Diamond v. Chakrabarty, 447 U.S. 303 (1980).

7. See, for example, Judith P. Swazey, James R. Sorenson, and Cynthia B. Wong, "Risks and Benefits, Rights and Responsibilities: A History of the Recombinant DNA Research Controversy," *Southern California Law Review* 51 (1978), 1019–77; Sheldon Krimsky, *Genetic Alchemy* (Cambridge, Mass.: MIT Press, 1982); Donald S. Frederickson, "Asilomar and Recombinant DNA: The End of the Beginning," in Kathi E. Hanna, ed., *Biomedical Politics* (Washington, D.C.: National Academy Press, 1991); Susan Wright, *Molecular Politics* (Chicago: University of Chicago Press, 1994).

8. Robert Pollack, a cancer researcher at Columbia University, has described his reactions as follows: "I had a fit. SV40 is a small animal tumor virus; in tissue cultures in the lab, SV40 also transforms individual *human* cells, making them look very like tumor cells. And bacteriophage lambda just naturally lives in *E. coli,* and *E. coli* just naturally lives in people . . . And I said, of all stupid things, at least put it into a phage, then, that doesn't grow in a bug that grows in your gut!" Swazey, Sorenson, and Wong, "Risks and Benefits," p. 1021.

9. Michael Rogers, *Biohazard* (New York: Knopf, 1977), p. 48.

10. Anne L. Hiskes and Richard P. Hiskes, *Science, Technology, and Policy Decisions* (Boulder: Westview, 1986), p. 127.

11. Judith Areen, Patricia A. King, Steven Goldberg, and Alexander M. Capron, *Law, Science, and Medicine* (Mineola, N.Y.: Foundation Press, 1984), p. 46.

12. U.S. Congress, Office of Technology Assessment (OTA), *New Developments in Biotechnology—Background Paper: Public Perceptions of Biotechnology* (Washington, D.C.: U.S. Government Printing Office, 1987), p. 86.

13. However, 78 percent of those polled said they would undergo genetic therapy to cure a serious genetic disease. OTA, *New Developments in Biotechnology,* p. 75.

14. Swazey, Sorenson, and Wong, "Risks and Benefits," pp. 1068–73.

15. 447 U.S. 303 (1980).

16. 35 U.S.C. §101.

17. 35 U.S.C. §101 provides that "Whoever invents or discovers any new and useful process, machine, manufacture, or composition of matter, or any new and useful improvement thereof, may obtain a patent therefor, subject to the conditions and requirements of this title."

18. *Revision of Title 35, United States Code,* S. Rep. No. 1979, 82nd Cong., 2nd sess. (1952), p. 5.

19. 447 U.S. at 317.

20. 447 U.S. at 322.

21. Some at least in the genetic engineering community have disagreed with Brennan's assessment and shared the majority's skepticism about the importance of patent protection for the progress of biotechnology. Factors that could dilute the value of biotechnology patents include delays between filing and issuance of patents and the requirement that samples of the novel organism be deposited in a collection where they would be accessible to competitors. Areen, King, Goldberg, and Capron, *Law, Science, and Medicine,* p. 108.

22. The oyster had been genetically modified so as to remain sweet through its reproductive period, a time when the organism is normally inedible.

23. Keith Schneider, "House Panel Rebuffs Its Staff on Animal Patents," *New York Times,* March 31, 1988, p. A19.

24. An article in *Science* reporting on the revolutionary event carried a picture of Leder captioned "fortuitously, a benevolent first case." William Booth, "Animals of Invention," *Science* 240 (1988), 718. Given the years of delay between *Chakrabarty* and the Leder patent,

it seems more reasonable to conclude that the Patent Office considered the potential benefits for cancer patients very carefully indeed before taking this controversial step.

25. Economic interests troubled by the "Harvard mouse" patent included farmers and ranchers, who strongly disapproved of the possibility that they would have to pay royalties for the offspring of patented livestock. In the past, patent rights had generally been considered to be exhausted at the point of sale. In the case of genetically altered animals, however, application of this rule would permit the purchaser of the new, superior breed to sell the offspring, thereby entering into competition with the patent holder.

26. *Federal Register* 49, October 17, 1984, pp. 40659–61. EPA concluded that genetically engineered organisms were encompassed within TSCA's definition of "chemical substance."

27. House Committee on Science and Technology, Subcommittee on Investigations and Oversight, *Issues in the Federal Regulation of Biotechnology: From Research to Release* (hereafter cited as *Biotechnology Report*), 99th Cong., 2nd sess. (1984), p. 31.

28. Marjorie Sun, "EPA Suspends Biotech Permit," *Science* 232 (1986), 15.

29. Keith Schneider, "Biotech's Stalled Revolution," *New York Times Magazine,* November 16, 1988, p. 47.

30. FIFRA Scientific Advisory Panel Subpanel, "Review of the Agency's Scientific Assessment of the Monsanto Application for an Experimental Use Permit to Field Test a Genetically Engineered Microbial Pesticide," Washington, D.C., April 25, 1986, p. 2.

31. Schneider, "Biotech's Stalled Revolution."

32. *Biotechnology Report,* pp. 22–23.

33. Ibid., p. 48.

34. Mark Crawford, "NIH Finds Argentine Experiment Did Not Break U.S. Biotechnology Rules," *Science* 235 (1987), 276.

35. Philip Boffey, "Tree Expert Ignores Federal Rules in Test of Altered Bacteria," *New York Times,* August 14, 1987, p. A1.

36. As defined in the guidelines, recombinant DNA molecules are "either (i) molecules which are constructed outside living cells by joining natural or synthetic DNA segments to DNA molecules that can replicate in a living cell, or (ii) DNA molecules that result from the replication of those described in (i) above." Strobel's study used a recombinant plasmid, which, however, did not replicate fully in the host organism that Strobel "deliberately released" into the environment, a strain of *Pseudomonas syringae.* Hence the released organism did not contain an rDNA molecule in the technical sense. "Report of the Committee to Review Allegations of Violations of the National

Institutes of Health Guidelines for Research Involving Recombinant DNA Molecules in the Conduct of Studies Involving Injection of Altered Microbes into Elm Trees at Montana State University," Washington, D.C., December 15, 1987.

37. Schneider, "Biotech's Stalled Revolution," p. 66.

38. 447 F. Supp. 668 (D.D.C. 1978).

39. 587 F. Supp. 753 (D.D.C. 1984).

40. Foundation on Economic Trends v. Heckler, 756 F.2d 143 (D.C. Cir. 1985).

41. 756 F.2d at 149.

42. The NIH director approved field testing requests from researchers at Stanford University and Cornell University, respectively, in August 1981 and April 1983. Both experiments, however, had to be canceled because of feasibility problems. The field test proposed by Steven Lindow and Nickolas Panopoulos of the University of California at Berkeley thus became the first NIH-approved deliberate release experiment in June 1983.

43. In February 1984, for example, a congressional subcommittee report concluded that NIH's regulatory framework did not guarantee proper consideration of the hazards of deliberate release experiments. Foundation on Economic Trends v. Heckler, 756 F.2d at 150.

44. 587 F. Supp. at 759.

45. 587 F. Supp. at 760.

46. 756 F.2d at 153. The federal courts are especially skeptical about brief explanations from expert bodies when they concern risks that judges and the public view as potentially catastrophic. For example, the D.C. Circuit Court was dissatisfied with the EPA administrator's "one sentence discussion" of the carcinogenicity of aldrin and dieldrin in Environmental Defense Fund v. EPA, 465 F.2d 528, 537 (D.C. Cir. 1972).

47. Maxine Singer, "Genetics and the Law: A Scientist's View," *Yale Law and Policy Review* 3 (1985), 333.

48. Ibid., p. 332.

49. For an analogous case, see the failure of EPA's Scientific Advisory Panel to understand the concept of "emergency suspension" under the Federal Insecticide, Fungicide, and Rodenticide Act. Jasanoff, *The Fifth Branch*, chap. 8. Of course, it could be argued that Steven Schatzow was making a comparable mistake when he interpreted the notion of deliberate release in the AGS case in purely commonsensical terms (a tree is not a "contained facility"). Legal concepts relevant to the regulation of hazardous technologies often have a scientific component that cannot be intuited through common sense alone.

50. 587 F. Supp. at 761.

51. Adrienne B. Naumann, "Federal Regulation of Recombinant DNA Technology: Time for Change," *High Technology Law Journal* 61 (1986), 88.

52. Scott D. Deatherage, "Scientific Uncertainty in Regulating Deliberate Release of Genetically Engineered Organisms: Substantive Judicial Review and Institutional Alternatives," *Harvard Environmental Law Review* 11 (1987), 203–246.

53. It is questionable whether courts would in practice have distinguished the proposed new standard of review (the "presumption" standard) from the standards prescribed by the Administrative Procedure Act. Prior experience suggests that the way courts look at agency decisions at the frontiers of scientific knowledge are influenced more by the nature of the issues under consideration than by the precise legal wording of the standard of review. See, for example, Sheila Jasanoff, "The Problem of Rationality in U.S. Health and Safety Regulation," in Roger Smith and Brian Wynne, eds., *Expert Evidence: Interpreting Science in the Law* (London: Routledge, 1989), pp. 154–157.

54. See, for example, David Dickson, *The New Politics of Science* (New York: Pantheon Books, 1984).

55. Foundation on Economic Trends v. Weinberger, 610 F. Supp. 829 (D.D.C. 1985).

56. 637 F. Supp. 25 (D.D.C. 1986).

57. 25 ERC 1429 (D.D.C. 1986).

58. 817 F.2d 882 (D.C. Cir. 1987).

59. *Federal Register* 49, December 31, 1984, p. 50856.

60. *Federal Register* 51, June 26, 1986, pp. 23302–93.

61. The Coordinated Framework specifically rejected the possibility of new legislation on the grounds that this would lead to greater uncertainty and that the full range of products generated through genetic engineering could not easily be accommodated within a single statutory approach. Ibid., p. 23303.

62. For an illustration, see the attack on EPA's biotechnology policy by Henry Miller, a fellow at the conservative Hoover Institution, who as an FDA official had been one of the foremost regulatory opponents of the so-called process-based approach to regulation. Henry I. Miller, "A Need to Reinvent Biotechnology Regulation at EPA," *Science* 266 (1994), 1815–18.

63. One example is Virginia's Biotechnology Research Act of 1994, discussed in S. Brian Farmer and Brian L. Buniva, "Virginia's New Biotechnology Law: Guidance for an Emerging Industry," *Virginia Bar Association Journal* 20 (1994), 3–5. In a bow to social containment,

the law prohibits localities in the state from enacting their own ordinances regulating or prohibiting biotechnology activities.

64. Sheila Jasanoff, "Product, Process or Program: Three Cultures and the Regulation of Biotechnology," in Martin Bauer, ed., *Public Resistance to New Technologies* (Cambridge: Cambridge University Press, 1995), pp. 311–331. Judy J. Kim, "Out of the Lab and into the Field: Harmonization of Deliberate Release Regulations for Genetically Modified Organisms," *Fordham International Law Journal* 16 (1993), 1160–1207.

65. Edward Yoxen, *The Gene Business* (New York: Oxford University Press, 1983), p. 181.

8. Family Affairs

1. These innovations include imaging techniques, surgery, and prenatal testing. They create new choices and new conflicts; see, for instance, Gina Kolata, "Operating on the Unborn," *New York Times Magazine,* May 14, 1989, pp. 34–35, 46–48.

2. Wiebe E. Bijker, Thomas P. Hughes, and Trevor Pinch, eds., *The Social Construction of Technological Systems* (Cambridge, Mass.: MIT Press, 1987).

3. In referring to constitutional law as a "literary technology," I am extrapolating from the very productive examination of language as a technology for stabilizing knowledge claims in the history of science. See particularly Steven Shapin and Simon Schaffer, *Leviathan and the Air-Pump: Hobbes, Boyle, and the Experimental Life* (Princeton: Princeton University Press, 1985); also see Peter Dear, ed., *The Literary Structure of Scientific Argument: Historical Studies* (Philadelphia: University of Pennsylvania Press, 1991).

4. 410 U.S. 113 (1973).

5. Notable contributions to the literature on the case include John Hart Ely, "The Wages of Crying Wolf: A Comment on *Roe v. Wade,*" *Yale Law Journal* 82 (1973), 920–949; Donald H. Regan, "Rewriting *Roe v. Wade,*" *Michigan Law Review* 77 (1979), 1569–1646; Catharine MacKinnon, "Privacy v. Equality: Beyond *Roe v. Wade,*" in *Feminism Unmodified: Discourses on Life and Law* (Cambridge, Mass.: Harvard University Press, 1987); Laurence H. Tribe, *Abortion: The Clash of Absolutes* (New York: Norton, 1990).

6. For overviews of work on actor-network theory in the sociology of technology, see Bijker, Hughes, and Pinch, *The Social Construction of Technological Systems;* Michel Callon, "Four Models for the Dynam-

ics of Science," in Sheila Jasanoff, Gerald E. Markle, James C. Petersen, and Trevor Pinch, eds., *Handbook of Science and Technology Studies* (Thousand Oaks, Calif.: Sage Publications, 1995), pp. 29–63.

7. David J. Garrow, *Liberty and Sexuality: The Right to Privacy and the Making of Roe v. Wade* (New York: Macmillan, 1994).

8. 381 U.S. 479 (1965).

9. 60 U.S.L.W. 4795 (1992).

10. 381 U.S. at 486–486.

11. 478 U.S. 186 (1986).

12. 405 U.S. 438 (1972).

13. See Garrow, *Liberty and Sexuality*, p. 658.

14. See, for example, Pope Paul VI, Encyclical Humanae Vitae, *Acta Apostolicae Sedes* 60 (1968), 481–503; see also Doctrinal Statement, "Instruction on Respect for Human Life in Its Origin and on the Dignity of Procreation: Replies to Certain Questions of the Day," March 10, 1987.

15. The third trimester is defined as the period after the fetus becomes viable or "potentially able to live outside the mother's womb, albeit with artificial aid." 410 U.S. at 160.

16. Garrow, *Liberty and Sexuality*, p. 599.

17. 462 U.S. 416, 458 (1983).

18. See, for example, Nancy K. Rhoden, "Trimesters and Technology: Revamping *Roe v. Wade*," *Yale Law Journal* 95 (1986), 639–697.

19. Garrow, *Liberty and Sexuality*, pp. 582–584.

20. 492 U.S. 490 (1989).

21. 60 U.S.L.W. at 4806–07.

22. State-mandated viability testing for fetuses above twenty weeks was a feature of the state law upheld in *Webster v. Reproductive Health Services.*

23. In Benten v. Kessler, 112 S.Ct. 2929 (1992), the Supreme Court upheld the ban on administrative law grounds. For an account of the events that led to the FDA ban, see R. Alta Charo, "A Political History of RU-486," in Kathi E. Hanna, ed., *Biomedical Politics* (Washington, D.C.: National Academy Press, 1991), pp. 43–93.

24. Garrow, *Liberty and Sexuality*, p. 692.

25. 60 U.S.L.W. at 4801. More specifically, the opinion suggested that *Roe* had caused people to rely on "the availability of abortion in the event that contraception should fail." This reasoning, in my view, reverses the probable direction of causality. I have argued that *Roe* won wide adherence, though admittedly at great cost, because the ready availability of contraception, and resulting freedom not to initi-

ate pregnancy, had made many people count on at least limited freedom to terminate early-stage pregnancies without undue interference
from the state. In other words, *Roe* was at least as much a symptom
as an agent of fundamental social change.

26. Ethics Advisory Board, *Report and Conclusions: HEW Support of
Research Involving Human In Vitro Fertilization and Embryo Transfer*
(Washington, D.C.: U.S. Government Printing Office, 1979).

27. Peter Singer, "Technology and Procreation: How Far Should We
Go?" *Technology Review* 88 (February/March 1985), 29. See also Clif-
ford Grobstein, Michael Flower, and John Mendeloff, "External
Human Fertilization: An Evaluation of Policy," *Science* 222 (1983),
131; Gina Kolata, "Frozen Embryos: Few Rules in a Rapidly Growing
Field," *New York Times,* June 5, 1992, p. A10.

28. Lori B. Andrews, *New Conceptions: A Consumer's Guide to the
Newest Infertility Treatments, Including In Vitro Fertilization, Artificial
Insemination, and Surrogate Motherhood* (New York: St. Martin's
Press, 1984), p. 147.

29. 556 F. Supp. 157 (N.D. Ill. 1983).

30. 74 Civ. 3588 (S.D.N.Y. 1978). For a report of this decision, see
Bioethics Reporter (Court Cases), no. 1/2 (1985), 7–24.

31. 59 U.S.L.W. 2205 (Tenn. App. 1990).

32. Ellen Goodman, "The Law vs. New Fact of Life," *Boston Globe,*
January 26, 1995, p. 13.

33. Andrews, *New Conceptions,* p. 150.

34. The principal exception is spina bifida.

35. Certain genetically transmitted physical deformities can be prevented through early detection and treatment. An example is congenital dislocation of the hip, a condition occurring more commonly in
female than in male children. See Colin Bruce and Henry R. Cowell,
"The Prevention of Genetically Determined Orthopaedic Defects,"
Clinical Orthopaedics and Related Research 222 (1987), 85–90.

36. For an interesting sampling of parental attitudes concerning
the birth of a defective child, see Letters, *New York Times,* June 9,
1988, p. C12.

37. Bonbrest v. Kotz, 65 F. Supp. 138 (D.D.C. 1946).

38. 49 N.J. 22 (1967).

39. The term "wrongful birth" was introduced into the law by anal-
ogy with "wrongful death," the legal term traditionally used to refer to
a death caused by a defendant's tortious conduct.

40. Thomas DeWitt Rogers III, "Wrongful Life and Wrongful Birth:
Medical Malpractice in Genetic Counseling and Prenatal Testing,"
South Carolina Law Review 33 (1982), 749–752.

41. Damages for emotional injuries, by contrast, were denied in

some jurisdictions on the ground that "parents may yet experience a love that even an abnormality cannot fully dampen." Becker v. Schwartz, 46 N.Y.2d 401, 414–415 (1978). Rogers suggests that any potential emotional benefits derived by parents from the birth of a child should be used to offset their claims for mental suffering rather than to deny them altogether. "Wrongful Life and Wrongful Birth," pp. 751–752. The speculative nature of such claims, however, may pose a barrier to recovery.

42. For examples of such rationales, see Bonnie Steinbock, "The Logical Case for 'Wrongful Life,'" *Hastings Center Report* 16 (1986), 15–20; and Rogers, "Wrongful Life and Wrongful Birth."

43. 49 N.J. at 28.

44. 80 N.J. 421 (1979).

45. 97 N.J. 339 (1984).

46. Steinbock, "The Logical Case for 'Wrongful Life,'" p. 15.

47. See, for example, ibid., p. 19 (citing *Wall Street Journal* article on legislation to restrict wrongful-life actions).

48. Although a variety of professionals may engage in the practice of genetic counseling, it seems advisable to hold them all to the same standard of care, at least as long as they are performing interchangeable services. The medical malpractice model of liability has generally been viewed as the most appropriate one to apply to the field of genetic counseling. Alexander M. Capron, "Tort Liability in Genetic Counseling," *Columbia Law Review* 69 (1979), 621–625.

49. On the failure to ask about family genetic history, see, for example, Karlsons v. Guerinot, 394 N.Y.S.2d 933 (1977); Phillips v. United States, 508 F. Supp. 544 (D.S.C. 1981); Turpin v. Sortini, 31 Cal.3d 220 (1982). On not recommending amniocentesis, see Berman v. Allen, 80 N.J. 421 (1979). Finally, in Naccash v. Burger, 223 Va. 406 (1982), a physician was held liable for mislabeling a father's blood sample so that he was not diagnosed as being a Tay-Sachs carrier. As a result, a child was born with Tay-Sachs disease.

50. George Annas, "Is a Genetic Screening Test Ready When the Lawyers Say It Is?" *Hastings Center Report* 15 (1985), 16–18. Neural tube defects (e.g., anencephaly, spina bifida) occur in 1 to 2 of every 1,000 live births in the United States.

51. Ibid., p. 17.

52. Careful analysts of the principles underlying wrongful-birth and wrongful-life actions have rejected this presumption. The New York Court of Appeals, for example, expressly denied this proposition in Becker v. Schwartz, 46 N.Y.2d 401 (1978), overruling an earlier decision that recognized a child's "fundamental right . . . to be born as a whole, functioning human being" (Park v. Chessin, 400 N.Y.S.2d

110 [1977]). See also Steinbock, "The Logical Case for 'Wrongful Life,'" p. 19 (arguing that the real "wrong" to the child is "to be born with such serious handicaps that many very basic interests are doomed in advance").

53. For further discussion of these issues, see Daniel Kevles, *In the Name of Eugenics: Genetics and the Uses of Human Heredity* (Berkeley: University of California Press, 1985); Dorothy Nelkin and Laurence Tancredi, *Dangerous Diagnostics: The Social Power of Biological Information* (New York: Basic Books, 1989); Neil A. Holtzman, *Proceed with Caution: Predicting Genetic Risks in the Recombinant DNA Era* (Baltimore: Johns Hopkins University Press, 1989).

54. Marcia Chambers, "Dead Baby's Mother Faces Criminal Charges on Acts in Pregnancy," *New York Times*, October 9, 1986, p. A22.

55. The court determined that the statute did not cover maternal negligence during pregnancy. Marcia Chambers, "Case against Woman in Baby Death Dropped," *New York Times*, February 28, 1987, p. A32.

56. Veronika E. B. Kolder, Janet Gallagher, and Michael T. Parsons, "Court-Ordered Obstetrical Interventions," *New England Journal of Medicine* 316 (1987), 1192–96.

57. Ibid., p. 1195.

58. See, for instance, Tamar Lewin, "Courts Acting to Force Care of the Unborn," *New York Times*, November 23, 1987, p. A1.

59. George Annas, "Protecting the Liberty of Pregnant Patients," *New England Journal of Medicine* 316 (1987), 1213.

60. Lewin, "Courts Acting to Force Care," p. B10.

61. The irony and depreciating quality of this term have frequently been noted. See, for example, William Safire, "The Modifiers of Mother," *New York Times Magazine*, May 10, 1987, pp. 10–12. In biological terms, the "surrogate" is in most cases the real mother, since she is both the source of the ovum and the person who gives birth to the child. Embryo transfer techniques have created a class of surrogates who are not genetically related to the child gestating in their wombs and who have been denied maternal rights by some courts.

62. According to a 1979 survey, AID accounts for between 6,000 and 10,000 live births annually in the United States. Martin Curie-Cohen, Lesleigh Luttrell, and Sander Shapiro, "Current Practice of Artificial Insemination by Donor in the United States," *New England Journal of Medicine* 300 (1979), 585–590.

63. Doornbos v. Doornbos, 23 U.S.L.W. 2308 (Ill. Super. Ct., Cook County, Dec. 13 1954); Gursky v. Gursky, 242 N.Y.S.2d 406 (Sup. Ct. 1963).

64. People v. Sorensen, 68 Cal.2d 280 (1968).

65. Strnad v. Strnad, 78 N.Y.S.2d 390 (Sup. Ct. 1948).

66. In this bizarre controversy, a woman artificially impregnated herself with sperm donated by an unmarried male friend, allegedly so as to avoid premarital sexual intercourse. The sperm donor subsequently demanded visitation rights to the child born out of this unorthodox procreative effort. 152 N.J. Super. 160 (1977).

67. For charges that courts in AID controversies have operated with narrow and outmoded views of the family, see Frank Grad, "Legislative Responses to the New Biology: Limits and Possibilities," *UCLA Law Review* 15 (1968), 503; C. Thomas Dienes, "Artificial Insemination: Perspectives on Legal and Social Change," *Iowa Law Review* 54 (1968), 285; Commentary, "Artificial Insemination: Problems, Policies and Proposals," *Alabama Law Review* 26 (1973), 160–161.

68. Despite legal uncertainties surrounding "surrogacy" contracts, an estimated 600 babies had been born to surrogate mothers by 1986. Thomas Eaton, "Comparative Responses to Surrogate Motherhood," *Nebraska Law Review* 65 (1986), 690. Surrogacy contracts had been tested in a number of states before the Whitehead–Stern controversy. In Doe v. Kelley, 106 Mich. App. 169 (1981), the first U.S. surrogacy case on record, the Michigan Court of Appeals had held that no money could be paid in connection with the adoption of a child under a surrogacy agreement, although the adoption itself was permitted to go forward. The Kentucky Supreme Court held in 1986 that commercial surrogacy arrangements did not violate state statutes prohibiting the sale or purchase of children. Somewhat misleadingly, the court characterized surrogacy as a solution "offered by science" to the problem of infertility and argued that it was up to the legislature to deal with the social and ethical questions raised by such developments. Surrogate Parenting Associates v. Kentucky *ex rel.* Armstrong, 704 S.W.2d 209 (1986).

69. *In re* Baby M, 217 N.J. Super. 313, 372 (1987).

70. *In re* Baby M, 109 N.J. 396 (1988).

71. Under New Jersey law, the husband of a woman who conceives through AID enjoys full parental rights in relation to the offspring. Stern argued that the infertile wife should be granted similar rights to a child born through surrogacy.

72. For further discussion of the mixed feminist responses to *Baby M,* see Nadine Taub, "Sorting Through the Alternatives," *Berkeley Women's Law Journal* 4 (1989), 285–299.

73. 286 Cal. Rptr. 361 (Cal. App. 4th Dist. 1991).

74. George Annas, "Using Genes to Define Motherhood," *New England Journal of Medicine* 326 (1992), 417–420.

75. 286 Cal. Rptr. at 380.

76. "'Baby M' Decision Creates Flurry of Legislative Activity," 13 FLR 1295, April 21, 1987. Bills to regulate surrogacy arrangements had been considered in several states as early as 1983, but as of 1985 no state had either legalized or prohibited surrogacy, although about twenty states and the District of Columbia were still considering the issue. John J. Mandler, "Developing a Concept of the Modern 'Family': A Proposed Uniform Surrogate Parenthood Act," *Georgetown Law Journal* 73 (1985), 1288.

9. Definitions of Life and Death

1. 497 U.S. 261 (1990).

2. Cruzan's family eventually returned to court with more evidence about their daughter's wishes, and the attorney general of Missouri decided not to contest their claims. On December 14, 1990, a state court ruled that feeding could be discontinued, thus allowing Cruzan to die. Andrew H. Malcolm, "Missouri Family Renews Battle over Right to Die," *New York Times*, November 2, 1990, p. A14; "Right-to-Die Case Nearing a Finale," *New York Times*, December 7, 1990, p. A4.

3. See, for example, Jessica Mitford, *The American Way of Death* (New York: Simon and Schuster, 1963).

4. Some 80 percent of deaths in the United States occur in hospitals or long-term care institutions. President's Commission for the Study of Ethical Problems in Medicine and Biomedical and Behavioral Research, *Deciding to Forego [sic] Life-Sustaining Treatment* (hereafter cited as *Forgoing Life-Sustaining Treatment*) (Washington, D.C.: U.S. Government Printing Office, 1983), pp. 17–18.

5. Earlier technological innovations, such as the stethoscope and cardiogram, provided improved tests for ascertaining death but did not necessitate a change from the prevailing definition of death as the cessation of breathing and the heartbeat. David Lamb, *Death, Brain Death, and Ethics* (Albany: SUNY Press, 1985), pp. 11–12. The adoption of the brain death standard did not eliminate all ambiguity, as became apparent in cases in which pregnant brain-dead women were kept "alive" to protect the unborn fetus. See Michael B. Green and Daniel Wikler, "Brain Death and Personal Identity," *Philosophy and Public Affairs* 9 (1980), 106–132.

6. New York State Task Force on Life and the Law, "The Determination of Death," July 1986, p. 3.

7. Norman Fost, "Do the Right Thing: Samuel Linares and Defen-

sive Law," *Law, Medicine and Health Care* 17 (1989), 330. The threat of liability may nevertheless be real; see, for example, Barber v. Superior Court, 147 Cal. App.3d 1006 (Cal. Ct. App. 1983).

8. *In re* Quinlan, 70 N.J. 10 (1976).

9. Superintendent of Belchertown v. Saikewicz, 373 Mass. 728 (1977).

10. The President's Commission for the Study of Ethical Problems found it remarkable that the first case involving treatment for unconscious patients arose only in 1975. Evidence heard by the *Quinlan* court indicated that decisions about this sort of care had routinely been made in the past without formal court review. *Forgoing Life-Sustaining Treatment,* p. 155.

11. Of course, litigation provides a single, highly specialized way of characterizing a social phenomenon as a "problem" and leads to an equally specialized form of solution: a legal remedy that can be designed and administered by the courts. Litigation is by no means the only technique for identifying issues as public problems, although it is a technique that is especially favored in the United States. See Joseph Gusfield, *The Culture of Public Problems: Drinking-Driving and the Symbolic Order* (Chicago: University of Chicago Press, 1981).

12. This term was used by Robert Burt in *Taking Care of Strangers* (New York: Free Press, 1979) to refer to patients who are unable to communicate with their caregivers.

13. At the time of *Quinlan,* a report issued by an ad hoc committee of Harvard Medical School was widely accepted as a source of criteria for defining "brain death." The Harvard report proposed the following criteria for determining what it termed "irreversible coma": unreceptivity and unresponsivity, total lack of reflexes, and lack of spontaneous movement and breathing. Ad Hoc Committee of the Harvard Medical School to Examine the Definition of Brain Death, "A Definition of Irreversible Coma," *Journal of the American Medical Association* 205 (1968), 337. The view that being in a vegetative state is not equivalent to death is widely held by the medical profession. See, for example, Lamb, *Death, Brain Death, and Ethics,* pp. 109–112.

14. The significance of *Quinlan* can be judged partly from the fact that there are about 10,000 comatose patients in America who may be described as being in a "persistent vegetative state." "Top Maine Court Backs Right to Die," *New York Times,* December 6, 1987, p. 41.

15. 70 N.J. at 50.

16. Patients in a persistent vegetative state (PVS) may continue to breathe spontaneously and retain circulatory functions for months and even years. One of the longest such cases on record is that of Elaine Esposito, who lapsed into such a state in August 1941 and

stopped breathing thirty-seven years later, in November 1978. Lamb, *Death, Brain Death, and Ethics,* p. 6. Recovery of consciousness in PVS patients is rare, but not unknown, and is usually accompanied by severe and permanent physical damage. *Forgoing Life-Sustaining Treatment,* pp. 182–183.

17. Jane D. Hoyt, "Karen Ann Quinlan's Fate Might Have Been Different," *Minneapolis Star and Tribune,* August 9, 1985, p. 15A.

18. 373 Mass. at 759.

19. 373 Mass. at 752–753.

20. Burt, *Taking Care of Strangers,* pp. 157–158. The President's Commission also recommended that the "best interests" standard be used in cases involving patients whose likely decision is unknown. *Forgoing Life-Sustaining Treatment,* p. 136.

21. 52 N.Y.2d 363 (1981).

22. *Forgoing Life-Sustaining Treatment,* p. 160.

23. Burt, *Taking Care of Strangers,* p. 154.

24. 70 N.J. at 39.

25. For an explanation of this term as applied to technology, see Wiebe E. Bijker, Thomas P. Hughes, and Trevor Pinch, eds., *The Social Construction of Technological Systems* (Cambridge, Mass.: MIT Press, 1987), p. 27.

26. On this point, see George Annas, "Reconciling *Quinlan* and *Saikewicz:* Decision Making for the Terminally Ill Incompetent," *American Journal of Law and Medicine* 4 (1979), 384.

27. Judith Areen, "Death and Dying," *Biolaw* 12 (1986), 277.

28. Burt, *Taking Care of Strangers,* p. 156.

29. Ibid., p. 157.

30. George J. Annas, "Whose Space Is This Anyway?" *Hastings Center Report* 16 (1986), 24.

31. Bouvia v. County of Riverside, No. 159780 (Cal. Super. Ct. Dec. 16, 1983)

32. Jay Horning, "Bedridden Bouvia Still Strong-Willed," *St. Petersburg Times,* April 25, 1993, p. 10A. For an earlier account, see Myrna Oliver, "Bouvia Still Wants the Right to Die," *Los Angeles Times,* May 23, 1988, p. 14.

33. Bartling v. Superior Court, 209 Cal. Rptr. 220 (Cal. App. 2d Dist. 1984).

34. E. J. McMahon, "Judge Allows Aged Patient to Starve Herself to Death," *New York Law Journal,* June 16, 1987, p. 1.

35. "Patient Allowed to End Treatment," *New York Times,* January 27, 1987, p. B9.

36. Peter Applebome, "An Angry Man Fights to Die, Then Tests Life," *New York Times,* February 7, 1990, p. A5.

37. Daniel Callahan, "On Feeding the Dying," *Hastings Center Report* 13 (1983), 22.

38. George J. Annas, "Do Feeding Tubes Have More Rights than Patients?" *Hastings Center Report* 16 (1986), 28.

39. "Top Maine Court Backs Right to Die," p. 41.

40. Brophy v. New England Sinai Hospital, 398 Mass. 417 (1986).

41. *Forgoing Life-Sustaining Treatment,* p. 83.

42. Ibid., pp. 88–89.

43. *In re* Conroy, 98 N.J. 321 (1985).

44. 497 U.S. at 288.

45. As of 1985, the law with respect to this issue was clearly established in only a dozen states. See House Select Committee on Aging, *Dying with Dignity: Difficult Times, Difficult Choices,* 99th Cong., 1st sess. (1985), pp. 26–27 (testimony of Barbara Mishkin).

46. Society for the Right to Die, *Handbook of 1985 Living Will Laws* (New York: Society for Right to Die, 1986), p. 5.

47. Ibid., pp. 16–17.

48. Myrna Oliver, "Controlling the End: Right-to-Die Laws Take On New Life," *Los Angeles Times,* May 23, 1988, p. 1.

49. *Forgoing Life-Sustaining Treatment,* p. 139.

50. Malcolm, "Missouri Family Renews Battle," p. A14.

51. See, for example, *Forgoing Life-Sustaining Treatment,* pp. 145–47; Barbara Mishkin, "Decisions concerning the Terminally Ill: How to Protect Patients, Staff and the Hospital," *HealthSpan* 2 (1985), 20.

52. Delaware was the first state to include such a provision in its natural death act. *Forgoing Life-Sustaining Treatment,* p. 145.

53. Society for the Right to Die, *Handbook of 1985 Living Will Laws,* p. 7.

54. This approach contrasts unfavorably with the New Jersey Supreme Court's flexible position in *Conroy.* Under that ruling, evidence of the patient's desires may be ascertained from a variety of sources, including both written and oral statements made by the individual while still competent.

55. A stringent interpretation of "terminal condition" could limit the application of many natural death acts to patients on the edge of death, although such individuals and their families have least to gain from a discontinuance of treatment. The Virginia Supreme Court declared in 1986 that the definition of "terminal condition" under that state's natural death act does not require death to be only hours away; a prognosis that death will occur within a few months is sufficient. Hazelton v. Powhatan Nursing Home, Inc., Supreme Court of Virginia, No. 860814, September 2, 1986.

56. Bartling v. Superior Court, 163 Cal. App. 3d 186 (1984).

57. *Forgoing Life-Sustaining Treatment,* pp. 217–223. In judging whether a treatment is beneficial to the child, the Commission proposed a very strict standard that would "exclude consideration of the negative effects of an impaired child's life on other persons, including

parents, siblings, and society." Ibid., p. 219. This proposal is difficult to reconcile with the Commission's equally firm view that the decision to undertake life-sustaining treatment creates an obligation to "provide the continuing care that makes a reasonable range of life choices possible." Ibid., p. 228. Such care would include adoption and foster care services in cases in which the parents are unable to raise the child themselves.

58. This appears to be a fair description of Baby Doe's condition, although the facts about her brain and the prognosis for her mental development were disputed. See Richard Sherlock, *Preserving Life: Public Policy and the Life Not Worth Living* (Chicago: Loyola University Press, 1987), p. 3.

59. Weber v. Stony Brook Hospital, 60 N.Y.2d 208 (1983).

60. U.S. v. University Hospital, 729 F.2d 144 (2d Cir. 1984).

61. Amendments to the Child Abuse Prevention and Treatment and Adoption Reform Act provided the legislative vehicle.

62. Norman Fost, "Infant Care Review Committees in the Aftermath of Baby Doe," in Arthur L. Caplan, Robert H. Blank, and Janna C. Merrick, eds., *Compelled Compassion* (Totowa, N.J.: Humana Press, 1992), pp. 285–297; see also Fost, "Do the Right Thing."

63. Linda Greenhouse, "Court Order to Treat Baby Prompts a Debate on Ethics," *New York Times,* February 20, 1994 (anencephalic baby given respiratory support at mother's request); Gina Kolata, "Battle over a Baby's Future Raises Hard Ethical Issues," *New York Times,* December 27, 1994, p. A1 (brain-damaged infant with urinary and intestinal problems kept on life support at parents' request).

64. For a classic exposition of the importance and the implications of asserting these values in public policymaking, see Guido Calabresi and Philip Bobbitt, *Tragic Choices* (New York: Norton, 1978).

65. *Forgoing Life-Sustaining Treatment,* p. 161.

66. Fost, "Infant Care Review Committees." See also "Symposium— Hospital Ethics Committees and the Law," *Maryland Law Review* 50 (1991), 742–919.

10. Toward a More Reflective Alliance

1. Lawyers may find less reassuring than scientists the observation by two highly credentialed professionals that DNA typing *before* the era of standardization resulted in no wrongful convictions, although it was based on unacceptable and sometimes indefensible practices. Eric S. Lander and Bruce Budowle, "DNA Fingerprinting Dispute Laid to Rest," *Nature* 371 (1994), 735.

2. The best evidence in support of this point comes from studies of

highly polarized regulatory disputes. See, for example, David Collingridge and Colin Reeve, *Science Speaks to Power* (London: Pinter, 1986); and Sheila Jasanoff, *The Fifth Branch: Science Advisers as Policymakers* (Cambridge, Mass.: Harvard University Press, 1990).

3. Arguably, the epidemiological evidence in the Bendectin cases established a strong enough *negative* case for general causation that courts were justified in according near-zero relevance to physicians' testimony on specific causation. See Joseph Sanders, "From Science to Evidence: The Testimony on Causation in the Bendectin Cases," *Stanford Law Review* 46 (1993), 36–47.

4. For evidence of such conflicts see Adeline Gordon Levine, *Love Canal: Science, Politics, and People* (Lexington, Mass.: Lexington Books, 1982); Phil Brown, "Popular Epidemiology: Community Response to Toxic Waste-Induced Disease in Woburn, Massachusetts," *Science, Technology, and Human Values* 12 (1987), 78–85; Sheila Jasanoff, ed., *Learning from Disaster: Risk Management after Bhopal* (Philadelphia: University of Pennsylvania Press, 1994). In all these cases, epidemiological work done by or on behalf of the plaintiffs' groups challenged work done by elite and/or state-authorized scientific bodies.

5. Included here are boundaries between "experience" and "experiment," as in the horizontal gaze nystagmus cases discussed in Chapter 3, and between "scientist" and "technician." On the latter point, see Sheila Jasanoff, "Judicial Construction of New Scientific Evidence," in Paul T. Durbin, ed., *Critical Perspectives in Nonacademic Science and Engineering* (Bethlehem, Pa.: Lehigh University Press, 1991), pp. 225–228.

6. For a case study developing this argument in detail, see Brian Wynne, *Rationality and Ritual: The Windscale Inquiry and Nuclear Decisions in Britain* (Chalfont St. Giles: British Society for the History of Science, 1982). See also Roger Smith and Brian Wynne, eds., *Expert Evidence: Interpreting Science in the Law* (London: Routledge, 1989).

7. On Bendectin, see Sanders, "From Science to Evidence." A study of the federal courts, however, showed more liability verdicts in judge-tried cases in federal courts. Kevin M. Clermont and Theodore Eisenberg, "Trial by Jury or Judge: Transcending Empiricism," *Cornell Law Review* 77 (1993), 1124–77. Further, a review of the tort system as a whole showed a high degree of judge-jury agreement. See Michael J. Saks, "Do We Really Know Anything about the Behavior of the Tort Litigation System—and Why Not?" *University of Pennsylvania Law Review* 140 (1992), 1230–41.

8. These examples are discussed in Francisco J. Ayala and Bert Black, "Science and the Courts," *American Scientist* 81 (1993), 234–236. In their insistence on a unitary model of "good" scientific prac-

tice and their unquestioning acceptance of the concepts of "testabil-ity" and "falsifiability," Ayala and Black display almost complete dis-regard for findings in contemporary historical, political, and sociolog-ical studies of science. For sharply contrary views of how science often works in practice, see the essays in Sheila Jasanoff, Gerald E. Markle, James C. Petersen, and Trevor Pinch, eds., *Handbook of Science and Technology Studies* (Thousand Oaks, Calif.: Sage Publications, 1995).

9. Brian Wynne, "Establishing the Rules of Laws: Constructing Expert Authority," in Smith and Wynne, *Expert Evidence,* p. 30.

10. "The operation of the Parentage Act does not depend on what a group of doctors, however distinguished and learned in their field, think the law ought to be." Quoted in George Annas, "Using Genes to Define Motherhood," *New England Journal of Medicine* 326 (1992), 420.

11. The sociological dynamics of credibility and skepticism in sci-ence are brilliantly expounded in Steven Shapin, *A Social History of Truth* (Chicago: University of Chicago Press, 1994). At a less theoreti-cal level, studies of scientific fraud and misconduct have pointed out that science generally functions in an atmosphere of credulousness that impedes scientists' ability to detect deviance. See William Broad and Nicholas Wade, *Betrayers of the Truth* (London: Oxford University Press, 1982).

12. Robert K. Merton, "The Normative Structure of Science," re-printed in Merton, *The Sociology of Science* (Chicago: University of Chicago Press, 1973), pp. 67–278. For an early critique, see Michael J. Mulkay, "Norms and Ideology in Science," *Social Science Information* 15 (1976), 637–656.

13. These tendencies in American administrative decisionmaking are particularly striking if one contrasts U.S. practices with those of European countries. See Ronald Brickman, Sheila Jasanoff, and Thomas Ilgen, *Controlling Chemicals: The Politics of Regulation in Europe and the United States* (Ithaca: Cornell University Press, 1985); and Sheila Jasanoff, *Risk Management and Political Culture* (New York: Russell Sage Foundation, 1986).

14. On a scientist's-eye view of mass tort litigation, see Marcia Angell, "Do Breast Implants Cause Systemic Disease?" *New England Journal of Medicine* 330 (1994), 1748–49. For reservations by lawyers, see Marc Galanter, "The Transnational Traffic in Legal Remedies," pp. 135–144; and Tom Durkin and William L. F. Felstiner, "Bad Arithmetic: Disaster Litigation as Less than the Sum of Its Parts," pp. 158–179; both in Jasanoff, *Learning from Disaster.*

15. Peter H. Schuck, "Multi-Culturalism Redux: Science, Law and

Politics," *Yale Law and Policy Review* 11 (1993), 14–40. Schuck argues that my own use of terms such as "science" and "scientist" shows that I believe they possess "some core, objective meaning not altogether contingent upon the speaker's idiosyncratic conceptions." Ibid., p. 38, n. 116. He is, of course, partly right. I do not believe that the terminology of science (including the very word "science") is contingent at the level of individual speakers. The structuring phenomena of socialization and institutionalization may leave individual speakers or practitioners very little freedom to play with the meaning of scientific concepts. This does not mean that contingency has been removed, merely that it has been driven deeper into the "black box," as noted by many sociologists of science and technology. See, in particular, Bruno Latour and Steve Woolgar, *Laboratory Life* (Princeton: Princeton University Press, 1986); and Latour, *Science in Action* (Cambridge, Mass.: Harvard University Press, 1987). Of course, similar observations about contingency and black-boxing could be made about seemingly unambiguous legal concepts as well.

16. Arthur Kantrowitz, "Proposal for an Institution for Scientific Judgment," *Science* 156 (1967), 763–764.

17. Richard Posner, "Will the Federal Court of Appeals Survive until 1984? An Essay on Delegation and Specialization of Judicial Function," *Southern California Law Review* 56 (1983), 775–790; Stephen L. Carter, "Separatism and Skepticism," *Yale Law Journal* 92 (1983), 1334–41.

18. Federal Judicial Center, *Reference Manual on Scientific Evidence* (Washington, D.C., 1994).

19. Lander and Budowle, "DNA Fingerprinting," p. 737.

20. Steven Goldberg uses this term in *Culture Clash: Law and Science in America* (New York: New York University Press, 1994), pp. 103–111. As used by Goldberg, the term overlaps with better-known terms such as "expert," "visible scientist," and "science adviser." Goldberg provides no sociological analysis of the science counselor's role. Accordingly, it remains unclear why science counselors would be motivated to participate in the legal system, on what basis they would command respect, and what cognitive commitments they would bring to the legal process.

21. This is arguably what happened in the Alar case when Uniroyal successfully defended its product against EPA's finding that it was a carcinogen. Sheila Jasanoff, "EPA's Regulation of Daminozide: Unscrambling the Messages of Risk," *Science, Technology, and Human Values* 12 (1987), 116–124. More generally, the adversarial procedures employed by the Scientific Advisory Panel encouraged the parties to engage in aggressive and often pointless discrediting of each

other's evidence. See also Jasanoff, *The Fifth Branch,* and "Procedural Choices in Regulatory Science," *Risk—Issues in Health and Safety* 4 (1993), 143–160.

22. Justice Holmes evidently had a linguistic metaphor in mind when he wrote about the Puerto Rican legal system that "to one brought up within it, varying emphasis, tacit assumptions, unwritten practices, a thousand influences gained only from life, may give to the parts wholly new values that logic and grammar could never have got from the books." Diaz v. Gonzalez, 261 U.S. 102, 106 (1923).

23. Some have speculated that professional societies could play a useful role by providing courts with lists of qualified experts. In 1990–91 a task force of the American Association for the Advancement of Science (AAAS) and the American Bar Association (ABA) explored this issue with representatives of nine scientific and engineering societies. All were enthusiastic about the possibility of developing a more active strategy for assisting the courts. The task force recommended to the Carnegie Commission on Science, Technology, and Government that professional societies of scientists and engineers consider educational and other activities to encourage their members to provide technical assistance to the courts. AAAS-ABA National Conference of Lawyers and Scientists, Task Force on Enhancing the Availability of Reliable and Impartial Scientific and Technical Expertise to the Federal Courts, "Report to the Carnegie Commission on Science, Technology, and Government," 1991.

24. Interview with Robert M. Cook Deegan, Washington, D.C., May 19, 1989.

25. In 1987, for example, the New York State Task Force on Life and the Law proposed that individuals should be authorized by law to designate a "health care proxy" to make medical decisions on their behalf, should they become incompetent. In 1989 the task force made some modifications that won the approval of a major Catholic group. Philip S. Gutis, "Patient Proxy for Treatment Gains Backing," *New York Times,* May 30, 1989, p. B1.

26. Institute of Medicine, *Society's Choices: Social and Ethical Decision Making in Biomedicine* (Washington, D.C.: National Academy Press, 1995).

27. General Accounting Office, "Product Liability: Extent of 'Litigation Explosion' in Federal Courts Questioned," Washington, D.C., January 1988. Though not framed in constructivist language, Saks's analysis of some of the "data" on tort litigation makes an interesting start in the direction I have proposed. Saks, "Do We Really Know?" pp. 1154–68 (see particularly his deconstruction of a table of claims frequencies provided by George Priest of Yale Law School). Saks's own

use of quantitative data, however, is not "reflexive" in the sense familiar to social studies of science; he is inconsistent in questioning the premises underlying the large number of studies he surveys in this otherwise admirable review essay.

28. Stephen LaTour, Pauline Holden, Laurens Walker, and John Thibaut, "Procedure: Transnational Perspectives and Preferences," *Yale Law Journal* 86 (1976), 283.

29. See in particular Sheila Jasanoff, "American Exceptionalism and the Political Acknowledgment of Risk," *Daedalus* 119 (1990), 61–81; idem, "Acceptable Evidence in a Pluralistic Society," in Deborah G. Mayo and Rachelle D. Hollander, eds., *Acceptable Evidence: Science and Values in Risk Management* (New York: Oxford University Press, 1991), pp. 29–47.

Index